SCIENCE AND THE ENDANGERED SPECIES ACT

Committee on Scientific Issues in
the Endangered Species Act

Board on Environmental Studies and Toxicology

Commission on Life Sciences

National Research Council

NATIONAL ACADEMY PRESS
Washington, D.C.   1995

NATIONAL ACADEMY PRESS • 2101 Constitution Ave., N.W. • Washington, D.C. 20418

NOTICE: The project that is the subject of this report was approved by the Governing Board of the National Research Council, whose members are drawn from the councils of the National Academy of Sciences, the National Academy of Engineering, and the Institute of Medicine. The members of the committee responsible for the report were chosen for their special competences and with regard for appropriate balance.

This report has been reviewed by a group other than the authors according to procedures approved by a Report Review Committee consisting of members of the National Academy of Sciences, the National Academy of Engineering, and the Institute of Medicine.

The project was supported by the Department of the Interior under Contract No. 1448000992010.

Library of Congress Cataloging-in-Publication Data

Science and the Endangered Species Act / Committee on Scientific
    Issues in the Endangered Species Act, Board on Environmental Studies
    and Toxicology, Commission on Life Sciences.
        p.   cm
    Includes bibliographical references and index.
    ISBN 0-309-05291-2
    1. Endangered species—United States. 2. Endangered species—Law
and legislation—United States. 3. Habitat conservation—United
States. I. National Research Council (U.S.).   Committee on
Scientific Issues in the Endangered Species Act.
QH76.S38   1995
333.9516'0973—dc20   95-33322

Printed in the United States of America

The National Academy of Sciences is a private, nonprofit, self-perpetuating society of distinguished scholars engaged in scientific and engineering research, dedicated to the furtherance of science and technology and to their use for the general welfare. Upon the authority of the charter granted to it by the Congress in 1863, the Academy has a mandate that requires it to advise the federal government on scientific and technical matters. Dr. Bruce Alberts is president of the National Academy of Sciences.

The National Academy of Engineering was established in 1964, under the charter of the National Academy of Sciences, as a parallel organization of outstanding engineers. It is autonomous in its administration and in the selection of its members, sharing with the National Academy of Sciences the responsibility for advising the federal government. The National Academy of Engineering also sponsors engineering programs aimed at meeting national needs, encourages education and research, and recognizes the superior achievements of engineers. Dr. Harold Liebowitz is president of the National Academy of Engineering.

The Institute of Medicine was established in 1970 by the National Academy of Sciences to secure the services of eminent members of appropriate professions in the examination of policy matters pertaining to the health of the public. The Institute acts under the responsibility given to the National Academy of Sciences by its congressional charter to be an adviser to the federal government and, upon its own initiative, to identify issues of medical care, research, and education. Dr. Kenneth I. Shine is president of the Institute of Medicine.

The National Research Council was organized by the National Academy of Sciences in 1916 to associate the broad community of science and technology with the Academy's purposes of furthering knowledge and advising the federal government. Functioning in accordance with general policies determined by the Academy, the Council has become the principal operating agency of both the National Academy of Sciences and the National Academy of Engineering in providing services to the government, the public, and the scientific and engineering communities. The Council is administered jointly by both Academies and the Institute of Medicine. Dr. Bruce Alberts and Dr. Harold Liebowitz are chairman and vice chairman, respectively, of the National Research Council.

Preface

The Endangered Species Act (ESA) is an important legislative tool for the protection of threatened and endangered species in the United States. The ESA asserts a legal claim on behalf of those species in the United States to habitat that sometimes conflicts with competing management goals for both private and public lands. It is inevitable that these conflicts play out in the political arena. Our committee was asked to provide advice on scientific aspects of the ESA and to consider whether the act is "protecting endangered species and their habitats." We have endeavored to restrict our advice to the areas where science can better inform the public policy debate. The distinction between science and public policy is often fuzzy, because the possession of scientific knowledge and the implementation of that knowledge are so closely linked. Our goal in this report has been to explore and illuminate the knowledge side of the equation.

Since the original passage of the ESA in 1973, scientific knowledge has been anything but static. Our understanding of biological species, in terms of their genetic and phylogenetic integrity, has greatly expanded since 1973. A rich array of new experimental tools has been acquired from both genetics and computational biology during the past two decades, and these have helped to drive a revolution in the traditional sciences of taxonomy and systematics. At the same time, new theoretical constructs have been elaborated that have given greater depth to definitions of species.

Species are composed of systems of populations (metapopulations) that have both temporal and spatial dimensions. The temporal history of individual species and of the migrating continental land masses that contain terrestrial habitats is known in much greater detail today than in 1973. The earth is dynamic and contemporary biological diversity is the unique realization of this long history of change. The time scales involved in biological change are long relative to human generations and, as a consequence, it is easy for us to see the biological world as static. Nothing could be further from the truth. Modern biology reveals that species are reservoirs of unique genetic adaptations to multifaceted physical and biological environments. The accumulation of these diverse adaptations is the result of a shared evolutionary history that typically involves hundreds of thousands of years of genetic continuity. The extinction of a species constitutes the irreversible loss of a suite of unique genetic adaptations that have been acquired (much like interest) over a long history of investment.

Rates of extinction are uneven over geological time. Several episodes of major extinction are now recognized, including the Permian-Triassic event (245 million years ago), when approximately 65% of terrestrial species became extinct, and the Cretaceous-Tertiary event (65 million years ago), when approximately 90% of terrestrial and marine reptiles became extinct. When viewed on a global scale, the present era constitutes yet another major episode of biological extinction. In contrast to the past, however, the present cause of extinction is a single biological species that has become so successful and so exploitive that it threatens to destroy the very capital that is necessary for its own long-term survival. That single species—humankind—is capable of rational analysis and planning, so that it can influence its own long-term destiny.

The earth's non-human biota is crucial to humans' long-term survival. We depend on the photosynthetic capability of green plants for the oxygen that we breathe and for virtually all of our food and energy requirements. The ability of green plants to grow is in turn dependent on a fixed supply of nitrogen (nitrates and nitrites) that are largely the product of a specialized group of microorganisms (Rhizobia). Many of our modern drugs have been derived from biotic sources. The list of human dependencies on the complex web of biological species is virtually endless.

Habitat, the spatial dimension of species, is absolutely crucial to species survival. Habitat is the theater in which the network of interactions between the physical and biological worlds play out. The landscape theory of habitat emphasizes the heterogeneity, complexity, and dynamic character of the physical and biological environment. The metapopulations of species are distributed on this shifting mosaic. If these are the scientific realities, then how do we match science to wise habitat conservation?

The authors of the ESA recognized that species conservation must include strong provisions for habitat conservation. These provisions included a trigger (threatened or endangered status of a species) that caused certain legal prohibitions (jeopardy and taking restrictions). The law provides for the recovery of species through the designation of critical habitat and through the elaboration and implementation of recovery plans. During the 20-year evolution of the ESA, additional provisions have been added, including additional mechanisms for habitat conservation, and others aimed at the resolution of conflicts engendered by ESA prohibitions. The committee was not charged with reviewing how the ESA is implemented by various federal agencies and did not directly address this question. We do, however, have several recommendations that would help improve the administration of the ESA if they were adopted (see Chapters 4 and 10, for example).

In general our committee finds that there has been a good match between science and the ESA. There are, of course, points where the agreement between science and the ESA is poorer. These include lack of timely designation of endangered or threatened status and similarly timely removal from these categories when recovery goals have been achieved. Survival habitat should be identified and designated for protection if necessary when species are listed as endangered. We have been able to align the "distinct population segment" language of the ESA with our contemporary understanding of evolutionary units. We hope that such alignment helps to achieve Congress's intent that distinct population segments be listed only sparingly and on a sound scientific basis and thus reduces the danger that the ESA itself could be jeopardized by carrying that language to an absurd extreme.

The analytical tools to evaluate species health have been greatly developed in recent years. The emergence of extinction theory from population genetics and ecology, the combination of demography and genetics in population viability analysis and the extension of risk analyses into the realm of biological conservation promise to lead us to wiser allocations of effort in the future. The field of ecosystem management has also emerged as a significant field of applied biology, in part as a response to the need for a more global view of conservation imperatives. The rich growth of these areas of science has also illuminated areas where our knowledge is still inadequate. In response to the charges given our committee, we attempt to identify areas of critical scientific uncertainty.

To paraphrase the great 20th century ecologist G. E. Hutchinson, species are the actors in the ecosystem theater. To sustain a viable future for our descendants, we must find ways to preserve both species and ecosystems. The ESA is a critically important part of our efforts to conserve species and thereby conserve ecosystems. By virtue of the habitat restrictions that accompany endangered status, species that happen to share habitat with an endangered species gain a measure of protection. The 20-year history of the ESA has validated its focus on species endangerment. Species are objective entities that are easily recognized. Their health and needs can be assessed and sound scientific management plans can be implemented.

Despite this, the task of managing each of the vast multitude of species on a case-by-case basis is beyond human capabilities. This is further compounded by the fact that many species remain undescribed. A challenge for the future is to find more integrated mechanisms to sustain both species and ecosystems that do not depend on case-by-case management.

It was my great good fortune to work with a knowledgeable, effective, and collegial committee. The various chapters of this report are the product of much hard work and spirited debate. I want to express my deep gratitude to the committee—including H. Ronald Pulliam, who resigned from the committee when he assumed the directorship of the National Biological Service in May 1994—for their wisdom, patience, and cheerful acceptance of the tasks imposed by this project. On behalf of the committee, I thank Project Assistant Adriénne Davis for attending to our many needs. Staff Officer Patricia Peacock was a source of much practical experience in conservation policy and she was a diligent editor and critic. Project Director David Policansky contributed his vast experience in science policy, especially in the realm of conservation policy, to this project. David Policansky and Pat Peacock also wrote, rewrote and edited many sections of this report. They contributed greatly to the finished product. Finally, thanks to the many representatives of public agencies—especially the Fish and Wildlife Service—and private groups who made written and oral presentations to our committee. They added an essential dimension to our understanding of the complex issues that surround the ESA.

Michael T. Clegg
Chairman

Contents

SCIENCE AND THE ENDANGERED SPECIES ACT

Executive Summary

INTRODUCTION

Species extinctions have occurred since life has been on earth, but human activities are causing the loss of biological diversity at an accelerating rate. The current rate of extinctions is among the highest in the entire fossil record, and many scientists consider it to have reached crisis proportions. The 1973 Endangered Species Act (ESA) and its subsequent amendments are the latest in a long line of federal legislation designed to protect wildlife. The ESA is the broadest and most powerful law to provide protection for endangered species and their habitats. The economic and social costs of complying with the ESA have been controversial in some cases. Because of those controversies, and because the act is being considered for reauthorization, it has been receiving much attention recently. That attention led to the request for this study to be conducted by the National Research Council (NRC).

The ESA defines three crucial categories: "endangered" species, "threatened" species, and "critical" habitats. ("Subspecies" of plants and animals and "distinct population segments" of vertebrates can also qualify for protection as species under the ESA.) Endangered species and their critical habitats receive extremely strong protection; it is illegal to take any endangered species of animal (or plant in some circumstances) in the United States, its territorial waters, or the high seas. In addition to this direct prohibition, Section 7 of the act prohibits any federal action that will jeopardize the future of any endangered species, including any threat to designated critical habitat. The act also requires the secretaries of interior and commerce to use programs in their agencies in furtherance of the act and requires other agencies to "utilize their authorities in furtherance of the purposes of [the act] by carrying out programs for the conservation of endangered species and threatened species." The 1978 and later amendments to the ESA established a requirement for recovery plans to be prepared by the U.S. Fish and Wildlife Service (FWS) for inland species and by the National Marine Fisheries Service for marine species, unless the secretary "finds that they will not promote the conservation of the species." Those plans are required to include specific population goals, timetables, and estimated costs.

The strength of the ESA lies with its stringent mandates constraining the actions of private parties and public agencies. Once a species is listed as threatened or endangered, it becomes entitled to shelter under the act's protective umbrella, a far-reaching array of provisions. Critical habitat must be designated "to the maximum extent prudent and determinable" and recovery plans, designed to bring the species to the point where it no longer needs the act's protections, are required if they will promote the conservation of the species. Funds for habitat acquisition and cooperative state programs are authorized. Federal agencies must ensure that their actions are not likely to jeopardize the survival of listed species nor adversely modify their critical habitats. Agencies are also required to use their authorities to promote endangered species conservation.

In addition to the Section 7 prohibition of any federal action that is likely to jeopardize an endangered species or adversely modify or destroy its critical habitat, Section 9 prohibits the taking of an endangered species of fish or wildlife[1] (or, by regulation, of threatened species). Sections 7 and 9 are major sources of the act's power as well as numerous controversies. In particular, the prohibition against taking endangered species has raised concerns among private landowners because of its application to habitat: *taking* is fairly broadly defined in the ESA and even more broadly in some regulations. How broad the definition of taking in regulations recently was reviewed by the U.S. Supreme Court. The court's decision will be important in determining the future of some of the controversies about the taking prohibition.

As human activities continue to affect species populations and their habitats, two major questions arise concerning the ESA. First, the focus of this report: is the ESA soundly based in science as an effective method of

---

[1]Section 9 provides somewhat lesser protection to plants, making it unlawful to "remove or reduce to possession any such species from areas under Federal jurisdiction . . . or remove, cut, dig up, or damage or destroy any such species on any other area in knowing violation of any law or regulation of any state . . .".

protecting endangered species and their habitats? The second question—of great public importance, but not part of this committee's charge—concerns the desired public policy with respect to protecting endangered species and their habitats, i.e., what are the costs and benefits, and to what extent is the public willing to incur the costs?

THE PRESENT STUDY

In November of 1991, Senator Mark Hatfield, Representative Thomas Foley, and Representative Gerry Studds wrote to the chairman of the National Research Council requesting a study of "several issues related to the Endangered Species Act." The request focused on scientific matters related to the act. After receiving funding from the U.S. Fish and Wildlife Service in September 1992, the NRC's Board on Environmental Studies and Toxicology convened the Committee on Scientific Issues in the Endangered Species Act. The committee's membership includes expertise in ecology; systematics; population genetics; wildlife management; risk and decision analysis; the legal, legislative, and administrative history of the Endangered Species Act; economics; and the implementation of the ESA from public and private perspectives. The committee's statement of task is based very closely on the letter of request from the three members of Congress (see Appendix A).

The committee was asked to review the following issues and to evaluate how they relate to the overall purposes of the Endangered Species Act:

• **Definition of species**. The committee was asked to review how the term *species* has been used to implement the ESA, and what taxonomic units would best serve the purposes of the act.

• **Conservation conflicts between species**. The committee was asked how frequent or severe conflicting conservation needs are when more than one species in a geographic area are listed as endangered or threatened under the ESA, and to make recommendations to resolve these conflicts.

• **Role of habitat conservation**. The committee was asked to evaluate the role of habitat protection in the conservation of species and to review the relationship between habitat-protection and other requirements of the act.

• **Recovery planning**. The committee was asked to review the role of recovery planning under the act and to consider how recovery planning could better contribute to the purposes of the act.

• **Risk**. The committee was asked to review the role of risk in decisions made under the ESA (such as what constitutes sufficient "endangerment" to require listing of a species, what constitutes jeopardy, adverse modifications, reasonable and prudent alternatives, taking, conservation, and recovery). It was also asked to review whether different degrees of risk ought to apply to different types of decisions (e.g., should an endangered species be at greater risk than a threatened species to justify listing?) and to identify practical methods for assessing risk to achieve the purposes of the act better while providing flexibility in appropriate circumstances to accommodate other objectives as well.

• **Issues of timing.** The committee was asked to review the timing of key decisions under the ESA and to consider ways of improving such timing under the act to serve its purposes better while minimizing unintended consequences.

The committee held meetings in Washington, D.C., and Irvine, California, where it received briefings and materials from federal officials, congressional staff, Secretary of the Interior Bruce Babbitt, members of private conservation organizations and of private industry, and other experts. It has also made use of many sources of information, including previous NRC reports; documents and studies done by other agencies; and relevant published literature from scientific journals, symposia, and books.

This report reviews scientific issues related to the ESA. The overall conclusion is that the ESA is based on sound scientific principles. Many scientific advances have been made since the ESA was passed in 1973, and they provide opportunities to improve the act's implementation, especially with respect to identifying species, subspecies, and distinct population segments, with respect to estimating risks of extinction, and with respect to economic and decision analyses. Although it is difficult to quantify the effectiveness of the act in preventing species extinction, there is no doubt that it has prevented the extinction of some species and slowed the declines of others. It is equally clear that the ESA by itself cannot prevent the loss of many species and their habitats. Instead, the ESA is best viewed as one part of a comprehensive set of ways of protecting species and their habitats. The committee was not asked to comment on the social and political decisions concerning the ESA's goals and tradeoffs, and it has not done so. Nonetheless, they are and should be an important part of the policy discussions about the ESA.

EXTINCTIONS

Extinction is an essential part of evolution. In the past 20 years, we have learned a great deal about the earth's physical and biological history. Over the past 500 million years, at least five mass extinctions have occurred, with as much as 84% of the genera of marine invertebrates disappearing from the fossil record. Those extinctions were associated with major physical events. Today, we are again witnessing a major extinction.

Unlike the earlier ones, which affected some kinds of organisms and some kinds of habitats more severely than others, today's extinctions are affecting all major groups of organisms in all nonmarine habitat types (the marine environment has not yet been affected as much as terrestrial and freshwater environments).

We do not know how many species of organisms live on earth, but there are many ways of estimating the rate of extinction in various habitats and in various kinds of organisms. The major cause of the current extinctions is human activity, and most estimates suggest that human activity has significantly increased the background extinction rate,[2] perhaps by orders of magnitude. Such activities include direct alteration of habitats by forestry, agriculture, fishing, and residential and commercial development; indirect alteration of habitats by pollution of water, air, and the soil; alteration of ecosystems by introductions of exotic organisms and the spread of diseases; removal or alteration of sources of food and shelter for organisms by human use of natural resources, and unregulated harvesting, hunting, and fishing.

THE SPECIES CONCEPT

Species of organisms are fundamental objects of attention in all societies, and different cultures have extensive literatures on the history of species concepts. The Endangered Species Act defines species to include "any subspecies of fish or wildlife or plants, and any distinct population segment of any species of vertebrate fish or wildlife which interbreeds when mature." In the act, the term *species* is used in a legal sense to refer to any of these entities. In addressing its use in the ESA, one must remember, however, that *species* has vernacular, legal, and biological meanings.

Many societies have notions of kinds of organisms, usually organisms that are large and conspicuous or of economic importance. The term *species* can be applied to many of those kinds and can be accurate as a scientific and vernacular term, because the characteristics used to differentiate species can be the same in both cases. Largely for this reason, the question of what a species is has not been a major source of controversy in the implementation of the Endangered Species Act. Greater difficulties have arisen in deciding about populations or groups of organisms that are genetically, morphologically, or behaviorally distinct, but not distinct enough to merit the rank of species—i.e., subspecies, varieties, and distinct population segments.

---

[2]Although the number of documented extinctions might appear to be small compared with the number of species alive, it is the rate of extinctions that is important. Even the mass extinctions of the past took many thousands of years to occur; the current rate of extinctions appears to be comparable to the rates during those events.

In particular, questions have arisen about how to recognize distinct population segments. To help in identifying them, the committee introduces the concept of an evolutionary unit (EU).[3] An EU is a group of organisms that represents a segment of biological diversity that shares a common evolutionary lineage and contains the potential for a unique evolutionary future. Its uniqueness can be sought in several attributes, including morphology, behavior, physiology, and biochemistry. Because any specified group of organisms can be claimed to have a unique evolutionary future, a basic characteristic of an EU is that it is distinct from other EUs. In most cases, an EU will also occupy a particular geographical area. Most currently recognized species and subspecies are EUs.

Distinction implies an *independent evolutionary future*. Estimates of distinctiveness (i.e., circumscription of EUs) are based on genetic, molecular, behavioral, morphological, or ecological characteristics. Any single method will often be inadequate to identify an EU (that is, to provide compelling evidence of distinctiveness). The question of distinctiveness and the associated inference of an independent evolutionary future usually requires the careful integration of several lines of evidence.

Committee Conclusion. The ESA is clear that species and subspecies of "fish or wildlife or plants"—defined in the act to include all members of the plant and animal kingdoms—are eligible for protection. The ESA's inclusion of distinct population segments—i.e., taxa below the rank of subspecies—is soundly based on science.

Committee Recommendation. The committee concludes that the ESA's inclusion of species and subspecies is soundly justified by current scientific knowledge and should be retained. Often, competent systematists will be required to delineate subspecies, and sometimes species as well.

Committee Recommendation. To help provide scientific objectivity in identifying population segments, the concept of the evolutionary unit (EU) should be adopted. The EU is a segment of biological diversity that contains a potential for a unique evolutionary future. To clarify the analyses, identifying an EU should be separate from deciding whether it is in need of protection.

Committee Conclusion. The ESA explicitly covers species and subspecies of all plants and animals. As currently written, however, it covers

---

[3]Similar but not identical to the National Marine Fisheries Service's Evolutionary Significant Unit; see Chapter 3.

taxonomic units below the subspecies level (i.e., distinct population segments) only for vertebrate animals. There is no scientific reason (other than lack of knowledge) to exclude any EUs of nonvertebrate animals and plants from coverage under the ESA. Although the way organisms are divided into kingdoms has changed since the ESA was enacted in 1973, current scientific knowledge about how species concepts apply to these organisms does not lead us to recommend that coverage be extended to prokaryotes and most single-celled eukaryotes, such as yeasts.

Committee Conclusion. Application of the EU concept should not result in any substantial change in the application of conservation laws. We hope it will move decisions of eligibility for protection away from arguments only about taxonomic ranks and into a realm where more substantive views about the degree to which populations are evolutionarily significant and new techniques can be applied.

HABITAT

Habitat—the physical and biological setting in which organisms live and in which the other components of the environment are encountered—is a basic requirement of all living organisms. It embraces all components of a species' environment. The relationship, nationwide, between vanishing habitats and vanishing species is well documented. The ecological relationship is simple and fairly general: species diversity is positively correlated with habitat area. A corollary of this relationship is that if habitat is substantially reduced in area or degraded, species occurring in the wild will be lost. Therefore, habitat protection is a prerequisite for conservation of biological diversity and protection of endangered and threatened species. The Endangered Species Act, in emphasizing habitat, reflects the current scientific understanding of the crucial biological role that habitat plays for species.

The question has been raised whether critical habitat should be determined at the time of listing or whether it should be deferred to the time of recovery planning. Because of public concern over economic consequences, the designation of critical habitat is often controversial and arduous, delaying or preventing the protection it was intended to afford.

Committee Recommendation. Because habitat plays such an important biological role in endangered species survival, some core amount of essential habitat should be designated for protection at the time of listing a species as endangered as an emergency, stop-gap measure. As discussed below, it should be identified without reference to economic impact. Economic review may need to remain linked to critical habitat determination in the ESA, and determination of areas essential to the recovery of a species, including areas not currently occupied by that species, can be especially complex. Hence we suggest designation of survival habitat.

Survival habitat would be designated at the time of listing of an endangered species, unless insufficient information were available or harm to the species would occur. For this purpose, survival habitat would mean the habitat necessary to support either current populations of a species or populations that are necessary to ensure short-term (25-50 years) survival, whichever is larger; survival habitat would receive the full protection that the ESA accords to critical habitat. Because of its emergency nature, no economic evaluation would be conducted before designating survival habitat. The designation of survival habitat (and its protection under the ESA) would automatically expire with the adoption of a recovery plan and the formal designation of critical habitat. Subsequent recovery planning would include designation of critical habitat as currently defined in the ESA (including economic evaluation) to include areas necessary for species recovery.

Because essential survival habitat is identified in our recommendation without reference to economic impact, and because it might not be sufficient to ensure long-term survival and recovery of endangered species, the committee views it as an emergency, stop-gap measure until critical habitat can be designated and a recovery plan can be completed, not as a substitute for those measures. Indefinite delays in designating critical habitat and formulating recovery plans after designation of survival habitat might cause harm to economic interests and to the endangered species itself. Therefore, implementation of this recommendation needs to include ways of preventing that delay from occurring.

Committee Recommendation. The committee endorses regionally based, negotiated approaches to the development of habitat conservation plans. Guidance from FWS for the development of such plans should include advice on the development of biological data, such as demographic and genetic analyses, habitat requirements of the species involved, reserve design, and monitoring, and it should also include advice on descriptions of management options and application of risk analyses in consideration of alternatives.

RECOVERY

The ultimate goal of the ESA is to recover threatened and endangered species. Recovery is "the process by which the decline of a threatened or endangered species is arrested or reversed, and threats to its survival are neutralized, so that its long-term survival in nature can be ensured." Despite increased attention from Congress, recovery plans are developed too slowly, and recovery planning remains handicapped by delays in its implementation, goals that are sometimes not scientifically supported, and the uncertainty of its application to other federal activities.

No recovery plan, however good it might be, will help prevent extinction or promote recovery if it is not implemented expeditiously. Indeed, the failure to implement a recovery plan quickly can also increase the disruption of human activities, because of the resulting uncertainty among other causes.

Committee Recommendation. To reduce uncertainty and permit the planning of activities not directed at species recovery, all recovery planning should include an element of "recovery plan guidance," particularly with regard to activities anticipated to be reviewed under sections 7, 9, and 10 of the ESA. FWS should convene a working group to develop explicit guidelines for the application of data to the construction of recovery objectives and criteria. To the degree possible, the guidance should identify activities that can be assumed to be consistent with the requirements of those sections, activities that can be assumed to be inconsistent with them, and activities that require individual evaluation. Topics would include a habitat-based approach to recovery; a logical, hierarchical approach to analysis of ecological and genetic data on the species; guidance for demographic modeling, stressing the inherent uncertainty of such modeling; outlining future research needs and how the research will contribute to species and habitat management; and an effective monitoring scheme.

Several habitat-related features of the ESA differ without scientific basis, in particular, standards applicable to the protection of plants and to the determination of jeopardy and modification of critical habitat, and different standards of protection on public and private lands. For example, Section 9 fails to protect endangered plants from habitat modification to the same degree that it protects animals, especially on private lands.

Committee Conclusion. The biological differences between animals and plants underlying their taxonomic separation offer no scientific reason for lesser protection of plants. The biological and physical requirements of species—including endangered and threatened species—do not vary according to the ownership of the habitats that they occupy. Therefore, there is no *biological* reason to have different standards for determination of "jeopardy," "survival," or "recovery" on public and on private lands (there could of course be other kinds of reasons).

Committee Conclusion. Public agencies and individual public servants on public lands behave differently from private landowners, both corporations and individuals, on private lands, because their rewards and incentives are different. Therefore, requirements applied equally on private and public lands will not necessarily provide the same degree of protection, although the *biological* standards or criteria on which the regulations are based are the same. It follows, then, that different mechanisms may be needed for avoiding endangerment and achieving recovery on public and private lands.

Committee Conclusion. The act and its regulations distinguish between species "survival" and "recovery" for purposes of determining jeopardy to species and adverse modification of their critical habitats. Survival and recovery are points on a continuum. Clearly, if a species does not survive, it cannot recover. It is less obvious, but still true, that any action that jeopardizes recovery also decreases the probability of long-term survival.

Committee Recommendation. To permit a rational evaluation of survival and recovery goals, estimates should be provided of probabilities of achieving various goals over various periods. The periods should be expressed both in years and in generation times of the organism of concern. Evaluation of long-term and irreversible impacts should be conducted in terms of long-term recovery of the species. Although it will often be difficult to make these estimates, even the attempt to make them will have value by requiring an objective analysis and by requiring assumptions to be specified.

CONSERVATION CONFLICTS BETWEEN SPECIES

Because plants and animals are linked to other organisms in ecosystems in a variety of ways, it is inevitable that conflicts will arise when attempts are made to protect individual species of plants or animals. One of the charges presented to the committee concerned conservation conflicts between species.

Committee Conclusion. We have found few well-documented cases where management practices focusing on particular species protected under the Endangered Species Act result in direct conflict with the needs of another.

It is possible that this low number stems from lack of knowledge of the ecological networks of which threatened and endangered species are part; from the fact that comparatively few species are currently listed and that recovery plans have been formulated for even fewer; and from the inadvertent protection for other listed species under some current recovery plans. We expect that our knowledge of such conflicts and the potential for their occurrence will increase as ecologies of listed species become better known, more recovery plans are formulated, and habitat for conserving endangered species becomes more constricted.

Committee Conclusion. Under current policies, the greatest potential for conflicts in protecting species and for management of individual species will arise in situations in which habitat reductions—especially extreme reductions—themselves are the causes of endangerment and the habitats of listed species are largely overlapping.

Committee Conclusion. The most effective way to avoid conflicts resulting from management plans for individual species is to maintain large enough protected areas to allow the existence of mosaics of habitats and dynamic processes of change within these areas. In addition to, and as part of, this strategy, multispecies plans should be devised to ensure the maintenance of habitat mosaics and ecological networks. Habitat (in the broadest sense) thus plays a crucial role in protecting individual target species and, ultimately, in reducing the need for listing additional species. When insufficient habitat is available to resolve such conflicts, other factors must be evaluated to resolve the conflicts, such as the consequences of various management options on each species, the ecological importance of the species, and the distribution of the species.

ESTIMATING RISK

The concept of risk is central to the implementation of the ESA. The main risks involved in the implementation of the Endangered Species Act are the risk of extinction (related to the probability of both biological and nonbiological events) and the risks associated with unnecessary expenditures or curtailment of land use in the face of substantial uncertainties about the accuracy of estimated risks of extinction and about future events. Since the passage of the ESA, there have been enough developments in conservation biology, population genetics, and ecological theory that substantially more scientific input can now be used in the listing and recovery-planning processes. Numerous models have been developed for estimating the risk of extinction for small populations. Although most of these models have shortcomings, they do provide valuable insights into the potential impacts of various management (or other) activities and of recovery plans. In particular, they are valuable for comparing the likely effects of alternative management options and of alternative adverse effects on the species.

Despite the major advances that have been made in models for predicting mean extinction times, the existing methods still have substantial limitations. Often, risk factors are not well known. Most of the models deal with only one risk factor at a time and fail to incorporate the interactive effects of multiple risk factors on reducing the time to extinction. This might result in a tendency for such models to underestimate the risk of extinction. Efforts to integrate various sources of random variation (genetic, demographic, and environmental) into spatially explicit frameworks are badly needed.

Most extinction models primarily address the mean time to extinction. Because decisions associated with endangered species usually are couched in fairly short time frames—less than 100 years—models that predict the cumulative probability of extinction through various time horizons would have greater practical utility than current models.

Committee Conclusion. With only a few exceptions, biologically explicit, quantitative models for risk assessment have played only a minor role in decisions associated with the ESA. They should play a more central role, especially as guides to research and as tools for comparing the probable effects of various environmental and management scenarios.

Committee Conclusion. Results from population-genetic theory provide the basis for one fairly rigorous conclusion. Small population sizes usually lead to the loss of genetic variation, especially if the populations remain small for long periods. If the members of the population do not mate with each other at random (the case for most natural populations), then the effect of small size on loss of genetic variation is made more severe; the population is said to have a smaller *effective size* than its true size. Populations with long-term mean sizes greater than approximately 1,000 breeding adults can be viewed as genetically secure; any further increase in size would be unlikely to increase the amount of adaptive variation in a population. If the effective population size is substantially smaller than actual population size, this conclusion can translate into a goal for survival for many species of maintaining populations with more than a thousand mature individuals per generation, perhaps several thousand in some cases. An appropriate, specific estimate of the number of individuals needed for long-term survival of any particular population must be based on knowledge of the population's breeding structure and ecology. If information on that species is lacking, information about a related species might be useful.

MAKING ESA DECISIONS IN THE FACE OF UNCERTAINTY

To ensure that ESA decisions protect endangered species as they are intended to in a scientifically defensible way requires objective methods for assessing risk of extinction and for assigning species to categories of protection according to that risk. Standards for assigning species to categories should be quantitative wherever possible and, when this is not possible, qualitative procedures should at least be systematic and clearly defined. Major advances in both theory and methods of estimating risk of extinction allow us to base listing and recovery decisions on scientific principles. In the past, many ESA decisions did not meet the guidelines suggested by current scientific thinking, listing species as endangered only when populations had dropped to the point where extinction was imminent and proposing recovery goals that left the species still at high risk of extinction.

Committee Conclusion. We can find no scientific basis for setting different levels of risk for different taxonomic groups, such as plants or animals, or for public versus private actions that may affect listed species. However, it is critical to understand that because public and private entities may behave differently, different management policies may be required for public and private lands in order to achieve the same *biological* risks for listed species in the two settings. No implementation of the ESA can be fully successful without recognizing these differences.

Committee Recommendation. To the degree that they can be quantified, the levels of risk associated with endangered status should be higher than those for threatened status. Once a species no longer qualifies for threatened status, it should be considered recovered and delisted. Levels of risk to trigger ESA decisions should be framed as a probability of extinction during a specified period (i.e., $x\%$ probability of extinction over the next y years). Although some crises may call for short time horizons (on the order of tens of years), ordinarily it will be necessary to view extinction over longer periods (on the order of hundreds of years) so that short-term solutions do not create long-term problems. The selection of particular degrees of risk associated with particular periods as the standards for listing species as endangered or threatened reflects both scientific knowledge and societal values.

Although the objectives of the ESA are not intrinsically conflicting, the act must be implemented with limited budgets, and so conflicts can arise in determining how to allocate funds among listed species, all of which qualify for the act's protection. Scientific considerations, such as whether a species or its habitat possesses unusually distinctive attributes or whether protection of a taxon would confer protection on other candidate taxa and their habitats, should be used to help set priorities for action. Decisions to set priorities for implementation of the act are often difficult and controversial, and the procedures for making them should be explicit and well documented. Structured methods, such as decision analysis, can improve both the substance of these decisions and the justifications offered for them.

Meeting the objectives of the act can sometimes conflict with other human objectives, such as development of private or public property harboring listed species. The act prohibits consideration of human objectives unrelated to species protection in decisions regarding listing, take, and jeopardy, but directs that these other objectives be taken into account in decisions about critical habitat and implementation of recovery plans. Tradeoffs between species protection and economic or other benefits or costs must be evaluated. Again, because these tradeoff decisions are often difficult and controversial, it is important to use well-structured and explicit methods for making them.

ESA decisions are inevitably based on limited information, and so agencies are obliged to act in the face of uncertainty about species status and the impacts of proposed activities. Decisions in the face of uncertainty carry the prospect of being wrong in various ways and with varying, and often asymmetrical, consequences. For example, managers concerned with delisting a formerly endangered species must be wary of two types of errors: delisting when the species is actually still in peril, and failing to delist when the species has truly recovered to the target level. Each type of error has both biological and nonbiological consequences. The first error has adverse biological consequences for the endangered species—it would be irreversible if the species became extinct—and, perhaps, positive socioeconomic consequences for sectors whose activities may have been constrained by recovery guidelines. The second error has neutral to positive consequences for the species but potential negative socioeconomic consequences. It is not possible to minimize the risks of both types of errors simultaneously. A decision rule that guards against the first will allow too many of the second and vice versa. To set acceptable rates for each type of error, both the likelihood and the magnitude of biological and nonbiological benefits and costs must be weighed in a decision-analytic framework. These decisions are too complicated and too consequential to be entrusted to unaided intuition.

If not examined explicitly, this asymmetric error structure can bias decisions under the act to the detriment of endangered species, especially if they are based on analyses that do not take the asymmetric risk function into account. Although the wording of the ESA suggests that the "burden of proof" to show no effect is on those proposing to modify habitat or harm a listed species, the way that hypothesis tests are phrased and error rates are set can put the burden on those attempting to show that a species should be listed or that a development proposal should be denied or modified.

Committee Recommendation. Because the structure of hypothesis testing related to listing and jeopardy decisions can make it more likely for an endangered species to be denied needed protection than for a nonendangered species to be protected unnecessarily, decisions under the act should be structured to take explicit account of all the types of errors that could be made and their consequences, both biological and nonbiological. The phrasing of the null hypothesis and setting of error rates should reflect societal, as well as scientific, judgments about what level of risk is acceptable for which types of errors.

TIMING

The committee's comments on the timing of key decisions under the ESA are incorporated in discussions of various other topics. In particular, timing is considered in discussions of recovery planning (where the committee concludes that recovery plans are developed too slowly and recovery planning remains handicapped by delays in implementation) and identification of survival habitat (whose designation is recommended to overcome the effects of delays in designation of critical habitat).

BEYOND THE ENDANGERED SPECIES ACT

The Endangered Species Act's goal is the prevention of species extinction, and its legal apparatus to protect endangered species is strong. It does not appear to have been intended as an overall policy act for the preservation of all of the nation's ecosystems and biota. It is, as the committee understands it, intended as a safety net.

Committee Conclusion. Although it is impossible to quantify the ESA's biological effects—i.e., how well it has prevented species from becoming extinct—the committee concludes that fewer species have become extinct than would have without the ESA. In other words, the ESA has successfully prevented some species from becoming extinct. Retention of the ESA would help to prevent species extinction. Some changes, as outlined in this report, would probably make the act more effective and predictable, and provide a more objective basis for its implementation.

Committee Conclusion. It is also clear that some species have become or are almost certain to become extinct despite the protection of the ESA. In other words, the ESA cannot *by itself* prevent all species extinctions, even if it is modified. Therefore, the committee concludes that additional approaches to the management of natural resources will need to be developed and implemented as complements to the ESA to prevent the continued, accelerating loss of species. Indeed, many federal, state, and local governments and private organizations are developing such approaches.
   • Ecosystem management. Despite diverse definitions of ecosystem management and despite scientific uncertainties, it is clear that managing ecosystems and landscapes as an addition to the protection of individual

species can lead to improved natural-resource management and can help reduce species extinctions. Properly implemented, it can also help to reduce uncertainty and thus reduce economic disruptions.

• Reconstruction or rehabilitation of ecosystems. Restoration ecology is a growing discipline. Many ecosystems functions have been improved or restored by such activities, and reconstruction or rehabilitation of ecosystem functioning holds much promise for the protection of endangered species. It is not usually possible to return an ecosystem to some prior pristine condition, however. Many ecosystems have been so altered that it is difficult to decide what prior condition we might want to return to. The trajectory taken by the ecosystem to get to its current condition is not retraceable in the way that a highway is, because many events occur in an ecosystem's history that are not precisely reversible. Genetic variability is lost; evolution occurs; exotic species are introduced; human populations in the region increase, and people develop dependence on a variety of modern technologies, cultures, and economic systems; and other natural and anthropogenic environmental changes affect the range of biophysical and socioeconomic possibilities for future states of the system. In brief, the past provides opportunities for the future but also constrains it. Thus, attempts to rehabilitate ecosystem functioning should keep these constraints in mind, so that inappropriately high expectations are not generated.

• Mixed management plans. Often, resource managers manage areas either for protection of biota or for human use. It is increasingly difficult to keep people and the effects of their activities separate from wildlife sanctuaries. Although such sanctuaries (e.g., national parks, wilderness areas, wildlife refuges, marine sanctuaries) are indispensable for protecting endangered species, greater attention needs to be paid to developing mixed-use areas. These would be urban recreation areas or residential and commercial developments adjacent to untrammeled areas designed to improve opportunities for wildlife while maintaining opportunities for human activities. Although the value of this approach is becoming increasingly recognized, its development is still in the early stages.

• Cooperative management. Various experiences with cooperative management—the sharing of planning and decision making by various government and nongovernment groups—have had some success. To some degree, habitat conservation plans represent an example of this approach, but it is likely that cooperative management will be necessary in cases where the strict requirements of the Endangered Species Act have not yet been applied. It is important to include the major interested parties without having so many interests involved that consensus is difficult to reach.

• Revised economic accounting. Too often, economic calculations underlying public and private decision making are incomplete. Often, they cover too short a time span, and they often exclude nonmarket values. A short-term loss might turn into a long-term gain: for example, losing an economic activity today might provide opportunities for greater economic activities of different types at some time in the future. Again, the validity of expanding economic accounting to cover longer periods and to include nonmarket values is becoming more widely recognized, but it is still in the early stages of development.

SCIENCE, POLICY, AND THE ESA

This committee was asked to review the scientific aspects of the ESA and it has done so. It has not uncovered any major scientific issue that seriously hinders the implementation of the act, although its review has suggested several scientific improvements. Many of the conflicts and disagreements about the ESA do not appear to be based on scientific issues. Instead, they appear to result because the act—in the committee's opinion designed as a safety net or act of last resort—is called into play when other policies and management strategies or their failures, or human activities in general, have led to the endangerment of species and populations. In some cases, policies and programs have been based on sound science, but other factors have prevented them from working. The committee does not see any likelihood that those endangerments will soon cease to occur or that the ESA can or should be expected to prevent them from occurring. It therefore concludes that any coherent, successful program to prevent species extinctions and to protect the nation's biological diversity is going to require more enlightened commitments on the part of all major parties to achieve success.

To conserve natural habitats, approaches must be developed that rely on cooperation and innovative procedures; examples provided for by the ESA are habitat conservation plans and natural community conservation planning. But those are only a beginning. Many other approaches have been discussed in various fora. They include cooperative management (sharing decision-making authority among several governmental and nongovernmental groups), transfer of development credits, mitigation banks, tax incentives, and conservation easements.

An analysis of these and other policy and management options is beyond this committee's charge, but sound science alone will not lead to successful prevention of many species extinctions, conservation of biological diversity, and reduced economic and social uncertainty and disruption. But sound science is an essential starting point. Combined with innovative and workable policies, it can help to solve these and related problems.

1

Introduction

Biological diversity—the variety of living things—has been of interest for centuries. In recent years, it has become apparent that human activities are causing the loss of biological diversity at an increasing rate: the current rate of extinctions appears to be among the highest in the fossil record. Although nonhuman organisms can cause extinctions of other species to a small degree, no other organisms produce such large effects over such wide areas as humans do and have done—at least locally—for thousands of years. Habitat alteration and degradation are probably the most severe effects humans have on other species today.

The Endangered Species Act (ESA) of 1973 (16 U.S.C. §§1531-1544 (1988)) is a far-reaching law that provides protection for threatened and endangered species and their habitats. Because of recent controversies and because the act is being considered for reauthorization, it has received much attention recently, which led to the request for this study by the National Research Council (NRC).

HISTORY

The ESA and its subsequent amendments are the latest in a long line of federal legislation designed to protect wildlife. Perhaps the first of the notable ancestors of the ESA was the Lacey Act of 1900 (Ch. 553 §1,31 Stat. 187 (1900) (codified as amended at 16 U.S.C. §701, 3371-3378 and 18 U.S.C. §42 (1988)). It prohibited interstate commerce of animals killed in violation of state game laws and stated that any dead animals taken across state boundaries were subject to the laws of the state into which they were imported; it also required the secretary of agriculture to take all measures to ensure the preservation, introduction, distribution, and restoration of game and wild birds. The Black Bass Act of 1926 (Ch. 346, 44 Stat. 576 (codified at 16 U.S.C. §§851-856 (1976) (repealed 1981)) was similar in concept. The Migratory Bird Treaty Act of 1918 (16 U.S.C. §§703-712 (1988)) prohibited the taking of certain birds protected by convention, and the Migratory Bird Conservation Act of 1929 (16 U.S.C. §§715-715s (1988)) provided acquisition authority for wildlife refuges and sanctuaries. The Endangered Species Preservation Act of 1966 (Pub. L. No. 89-669, 80 Stat. 926 (1966)) directed the secretary of the interior to conserve, protect, restore, and propagate selected species of native fish and wildlife; this act provided authority for the acquisition of land for habitat protection and charged the secretary to determine whether a species was threatened with extinction. In 1969, the Endangered Species Conservation Act (Pub. L. No. 91-135, 83 Stat. 275 (1989)) supplemented the Endangered Species Preservation Act by authorizing the secretary to promulgate a list of species or subspecies of fish and wildlife threatened with worldwide extinction and to prohibit their importation into the United States. It also amended and strengthened earlier acts.

In 1972, Congress passed the first of two broad, powerful acts that protect wildlife—the Marine Mammal Protection Act (16 U.S.C. §§1361-1407 (1988)). That act prohibited the taking or importation of any marine mammal, with exceptions available under certain complex conditions. It also provided for a determination of "depleted" status; depleted stocks or species were afforded stronger protection than others. And in 1973, Congress passed the ESA—the broadest and most powerful wildlife-protection act in U.S. history (Bean, 1983). The act defines three crucial categories: "endangered" species, "threatened" species, and "critical" habitats. ("Subspecies" of plants and animals and "distinct population segments" of vertebrates can also qualify for protection as species.) Endangered species and their critical habitats receive extremely strong protection; it is illegal to take any endangered species of animal, and plants in some circumstances, in the United States, its territorial waters, or the high seas. The term "to take" is defined as "to harass, harm, pursue, hunt, shoot, wound, kill, trap, capture, or collect, or attempt to engage in any such conduct." In addition to that direct prohibition, Section 7 of the act prohibits any federal action likely to jeopardize the future of any endangered species or to result in destruction or adverse modification of designated critical habitat. The act also requires the secretaries of interior and commerce to use programs in their agencies in furtherance of the act and requires other agencies to "utilize their authorities in furtherance of the purposes of this [act] by carrying out programs for the conservation of endangered species and threatened species." Amendments to the ESA in 1978 and later established the requirement for recovery plans to be prepared by the U.S. Fish and Wildlife Service for inland and certain marine species and by the National Marine Fisheries Service for other marine species and anadromous fishes. Those plans are required to include specific population goals, timetables, and estimated costs.

The power of the ESA lies with its stringent mandates constraining the actions of both private parties and public agencies. Once a species is listed as endangered or threatened in accordance with the ESA, it is sheltered under the act's protective umbrella, an impressive and far-reaching array of provisions (Greenwalt, 1988). Critical habitat must be designated unless overriding concerns for the species' conservation prevail (Section 4 of the ESA). Recovery plans designed to bring the species to the point where it no longer needs the act's protections are required unless they will not promote the conservation of the species (Section 4). Funds for habitat acquisition and cooperative state programs are authorized (sections 5 and 6). Federal agencies must ensure that their actions are not likely to jeopardize the continued existence of listed species nor adversely modify their critical habitats (Section 7(a)(2)). They are also required to use their authorities to promote endangered species conservation (Section 7(a)(1)).

Taking endangered species and, by regulation, threatened species, is prohibited with certain exceptions (Section 9). For example, listed plants are protected only on federal lands unless their taking on nonfederal lands violates state law. The act and regulations that implement it define taking very broadly to cover any activities causing harm to listed species, including habitat destruction if it interferes with behavior patterns essential to species survival and reproduction (however, the extent of this last provision will be decided by the Supreme Court, which has agreed to review a case on the matter[1]). Limited taking may be allowed under federal permits for research purposes. Taking that occurs incidentally to otherwise lawful actions of federal agencies is legal if the federal agency or its permittee complies with terms and conditions found in the biological opinion issued under Section 7(a)(2). Unless nonfederal actions trigger a Section 7 consultation, listed species may be taken only under authority of a Section 10(a)(1)(B) incidental take permit. To obtain the permit, an applicant must submit a habitat conservation plan, reduce and avoid take within feasible limits, mitigate for any unavoidable take, show capability to fund the plan's conservation measures, and meet any other conditions imposed by the secretary of Interior or of Commerce.

The Section 7 prohibition of any federal action likely to jeopardize the continued existence of an endangered or threatened species or to destroy or adversely modify its critical habitat is the source of much of the act's power.

---

[1]See Chapter 4.

It has led to controversies, the first of which concerned the Tellico Dam on the Little Tennessee River. That dam would have flooded all the known habitat of the newly discovered snail darter (*Percina tanasi*). The Supreme Court ruled 6-3 in 1978 (TVA v. Hill, 437 U.S. 153 (1978)) that Section 7 prohibited the closing of the gates of the dam. Chief Justice Burger wrote that one "would be hard pressed to find a statutory provision whose terms were any plainer than those in §7." Although Congress overruled itself in exempting the Tellico Dam from the provisions of the ESA, and other natural populations of the snail darter were subsequently found, an increasing number of controversies have arisen. A recent controversy, addressed by the National Research Council, involved sea turtles and shrimpers (NRC, 1990). A current controversy involves the northern subspecies of the spotted owl (*Strix occidentalis caurina*) and the timber industry in the Pacific Northwest. Another has the potential to intensify if more than 100 stocks of Pacific salmon (*Oncorhynchus* species) (see NRC, 1995) and the widespread but rare bull trout (*Salvelinus confluentus*) proposed for listing beome listed as endangered or threatened. (As of January 1995, four stocks of Pacific salmon were listed as endangered in the Pacific Northwest and California.) These fishes and their habitats are used for various economic, social, and recreational activities that adversely affect their survival in the western United States.

As human activities continue to affect species populations and their habitats, two major questions arise concerning the ESA. First, is the act soundly based in science as an effective method of protecting endangered species and their habitats? Congress intended the act to provide a mechanism for reversing the endangered or threatened status of species; that is why the processes of "delisting" species and upgrading them from endangered to threatened are in the act. The second question concerns the desired public policy with respect to protecting endangered species and their habitats, i.e., what are the costs and benefits, and to what extent is the public willing to incur the costs? This report focuses on the first question.

To address the first question, we return to the objectives of the ESA as described in the act. Section 2(b) states that the act's purposes are "to provide a means whereby the ecosystems upon which endangered species and threatened species depend may be conserved, to provide a program for the conservation of such endangered species and threatened species, and to take such steps as may be appropriate to achieve the purposes of the treaties and conventions set forth in subsection (a) of this section."

THE PRESENT STUDY

In November of 1991, Senator Mark Hatfield, Representative Thomas Foley, and Representative Gerry Studds wrote to the chairman of the National Research Council requesting a study of "several issues related to the Endangered Species Act," clearly focusing on scientific matters related to the act. After receiving funding from the U.S. Fish and Wildlife Service in June 1992, the NRC's Board on Environmental Studies and Toxicology convened the Committee on Scientific Issues in the Endangered Species Act. The committee's membership (see Appendix C) includes expertise in ecology; systematics; population genetics; wildlife management; risk and decision analysis; the legal, legislative, and administrative history of the Endangered Species Act; economics; and the implementation of the ESA from public and private perspectives. The committee's statement of task is based very closely on the letter of request from the three members of Congress (Appendix A).

The committee was asked to review the following issues and evaluate how they relate to the overall purposes of the Endangered Species Act:

• **Definition of species**. The committee was asked to review how the term *species* has been used to implement the ESA, and what taxonomic units would best serve the purposes of the act.

• **Conservation conflicts between species**. The committee was asked how frequent or severe conflicting conservation needs are when more than one species in a geographic area are listed as endangered or threatened under the ESA, and to make recommendations to resolve these conflicts.

• **Role of habitat conservation**. The committee was asked to evaluate the role of habitat protection in the conservation of species and to review the relationship between habitat-protection and other requirements of the act.

• **Recovery planning**. The committee was asked to review the role of recovery planning under the act and to consider how recovery planning could better contribute to the purposes of the act.

• **Risk**. The committee was asked to review the role of risk in decisions made under the ESA (such as what constitutes sufficient "endangerment" to require listing of a species, and what constitutes jeopardy, adverse modifications, reasonable and prudent alternatives, taking, conservation, and recovery). It was also asked to review whether different degrees of risk ought to apply to different types of decisions (e.g., should an endangered species be at greater risk than a threatened species to justify listing?) and to identify practical methods for assessing risk to achieve the purposes of the act better while providing flexibility in appropriate circumstances to accommodate other objectives as well.

• **Issues of timing**. The committee was asked to review the timing of key decisions under the ESA and to consider ways of improving timing under the act to serve its purposes better while minimizing otherwise unintended consequences.

The committee met in Washington, D.C., and Irvine, California, where it received briefings and materials from federal agencies, congressional staff, Secretary of the Interior Bruce Babbitt, members of private conservation organizations and of private industry, and other experts. It also made use of many sources of information, including previous NRC reports; government documents and studies done by other agencies; and relevant published literature from scientific journals, symposia, and books.

The report begins with a description of the processes of extinction as they have occurred through geological time to the present. This is followed by analyses of the scientific underpinnings of the ESA—the species concept, habitat and its role in species conservation, conservation conflicts between species, the assessment of extinction risks, decision-making under uncertainty, and those areas where lack of scientific knowledge or basic uncertainties need to be taken into account. Timing issues, mostly related to recovery and designation of critical habitat, are discussed in Chapter 4. These analyses provide the bases for the report's conclusions and recommendations.

REFERENCES

Bean, M.J. 1983. The Evolution of National Wildlife Law. 2nd Ed. New York: Praeger.

Greenawalt, L.A. 1988. Reflections on the power and potential of the Endangered Species Act. Endangered Species Update 5:7-9.

NRC (National Research Council). 1990. Decline of the Sea Turtles: Causes and Prevention. Washington, D.C.: National Academy Press.

NRC (National Research Council). 1995. Upstream: Salmon and Society in the Pacific Northwest. Washington, D.C.: National Academy Press.

2

Species Extinctions

EXTINCTIONS OVER GEOLOGICAL TIME

Any attempt to reduce human-caused species extinction (i.e., to protect endangered species) requires an understanding of extinction rates over geological time. Fossil records from nearly any geological age reveal two types of extinction: extinction without replacement ("dead-end") and what might be called "chronologic extinction" or "taxonomic extinction." The first type is genuine extinction; it is the end of an evolutionary lineage. The second type is based on the recognition by paleontologists that one species has changed through geological time (i.e., has evolved) to the extent that it is classified as a different species from the next earliest representative of its lineage. The species involved (see Martin and Barnosky, 1993; Nadachowski, 1993; Turner, 1993) represent an evolutionary continuum rather than an evolutionary dead-end. The discussion that follows is concerned only with dead-end extinction.

The past 20 years have witnessed major advances in our understanding of the earth's physical history (including understanding of paleoclimate, changing sea level, volcanism, and plate tectonics), and knowledge of past plant and animal communities is much improved. With refinements in radiometric dating, stratigraphy, and the discovery and analysis of varied sources of paleoenvironmental data (from cave deposits, ice cores, pollen cores, fossil dung, packrat middens, tree rings, etc.), we now can correlate many biotic and abiotic events in earth history with reasonable accuracy. What we learn from studying the past is important for predicting biotic responses to future changes in climate and other physical phenomena (Burney, 1993).

The marine invertebrate fossil record reveals at least five mass-extinction events during the past 500 million years, in which from 14% to 84% of the genera or families disappeared from the fossil record (Jablonski, 1991; Raup and Jablonski, 1993; Benton, 1995). Perhaps the two best known of these mass-extinction events are the Permian-Triassic (P-T) 245 million years ago and the Cretaceous-Tertiary (K-T) 65 million years ago. In addition to marine invertebrates, the P-T event involved extinction of terrestrial plants and insects (Retallack, 1995); the K-T event included most reptiles. A much more recent mass extinction, that of the Pleistocene-Holocene (P-H) only 11,000 years ago, featured extinction of more than 100 terrestrial species of birds and large mammals in North, Central, and South America. Although extinction is common to the P-T, K-T, and P-H events, they differ from the current extinction event in several ways.

The current extinction event differs from the three paleontological events in that it actually or potentially involves all groups of organisms, in any sort of nonmarine habitat. A particularly conspicuous aspect of today's situation is that many species of plants, as well as animals, are rare, endangered, or already have become extinct. For example, the Endangered Species List contained 695 U.S. and 521 non-U.S. endangered species and 169 U.S. and 41 non-U.S. threatened species in March 1995. Of that total, 995 were animals and 529 were plants.

In two of the other three extinction events, plants suffered few if any losses. The evolutionary radiation of mammals that followed the K-T demise of most reptiles was possible only because most terrestrial plants survived the K-T event. As far as can be determined from the extensive late Quaternary pollen and plant macrofossil record, the loss of so many birds and mammals in the Americas during the P-H event was not accompanied by the extinction of plants, although the geographic ranges of plants changed dramatically at that time (Betancourt et al., 1990; Webb et al., 1993).

Today, even though obscure or little-known species are lost with little or no notice (Wilson, 1992), we can document the date of loss of relatively well-known species to the nearest decade (e.g., the sea mink, *Mustela macrodon*, became extinct about 1880 (Nowak and Paradiso, 1983)), sometimes to the year (e.g., the Guam flycatcher, *Myiagra freycineti*, became extinct in 1985 (Engbring and Pratt, 1985; Steadman, 1992)), or even to the day, in the case of the last passenger pigeon, *Ectopistes migratorius*, which died on September 1, 1914 (Schorger, 1955). Such precise dating is not possible for the paleontological extinctions. Because they occurred within the time range of radiocarbon dating, the 95% confidence limits to the estimated dates of P-H extinctions are ± 50 to 200 years (Stafford et al., 1990). But for the K-T and P-T events, even the most precise radiometric dating methods have margins of error (a 95% confidence interval) of $\pm 10^4$ to 10^5 years and 10^5 to 10^6 years, respectively. Pinpointing prehistoric extinctions also depends upon the completeness of the fossil record, because failure to find a species in a certain fossil assemblage does not mean that the species is extinct; the absence could be a sampling artifact (Steadman et al., 1991).

The P-T and K-T events are not as well documented as the P-H event because they are represented by many fewer paleontological sites, and a much greater variety of organic materials have withstood the mere 11,000 years since the P-H event. Furthermore, most species that were lost in the P-T and K-T events are related only distantly to living species; in contrast, many birds and mammals lost in the P-H event are survived by closely related species or genera that can be studied to help determine the paleo-ecology of the extinct species.

A series of glacial-interglacial cycles have occurred during the past 2 million years. The cool glacial intervals have lasted 80,000 to 120,000 years, and the warm interglacial intervals have lasted only 10,000 to 20,000 years each. The temperature and atmospheric chemistry of our current interglacial period (the Holocene) have been unusually stable (Rampino and Self, 1992; Dansgaard et al., 1993; Mayewski et al., 1993); however, even subtle shifts in Holocene temperature or precipitation regimes can have dramatic local consequences on plant communities (Swetnam, 1993). The last interglacial interval, which probably had much more variable temperatures than the current one, lasted from about 130,000 to 115,000 years ago (Martinson et al., 1987; Shackleton, 1987; Appenzeller, 1993; Dansgaard et al., 1993; GRIP, 1993; Grootes et al., 1993). Global sea-level fluctuations between glacial and interglacial intervals have been on the order of 100 m (Shackleton, 1987; Pirazzoli, 1993).

Warm interglacial climates have characterized about 15% of the past 2 million years (Imbrie and Imbrie, 1979). During glacial intervals, which featured continental ice sheets, expanding alpine glaciers, and periodic "armadas" of icebergs across the North Atlantic (Kerr, 1993a), the average air, soil, and groundwater temperatures were much cooler than today, even in subtropical Florida (Plummer, 1993) and in truly tropical localities (Markgraf, 1989). In those times, species could survive the great continental climatic fluctuations because they could move gradually across intact habitats. The response of plants and animals to climatic changes of the next glacial interval will likely be impaired by habitat fragmentation. Species with poor dispersal capabilities might be unable to shift their ranges across tracts of disturbed habitat.

PREHISTORIC HUMAN IMPACT ON
CONTINENTAL ECOSYSTEMS

Although the causes of the P-T and K-T extinction events are topics of much interest, research, and speculation (e.g., Jablonski, 1991; Kerr, 1993b; Morell, 1993; Sharpton et al., 1993), we can be certain that humans were not involved. That might not be the case, however, for the P-H event and certainly is not so for the current extinction event. As far as can be determined, human activity has significantly increased rates of extinction—perhaps by orders of magnitude—over the background rate (see below) and therefore is the primary cause of the current extinction event (Wilson, 1992). How far back in time can we trace human impact?

Humans, including early forms of *Homo* and modern *Homo sapiens*, have occupied much of Africa and Eurasia for hundreds of millennia (Diamond, 1992; Klein, 1992). Archaeological evidence (especially bone assemblages and stone tools) suggests that people in Africa and Eurasia have hunted large mammals since at least 100,000 years ago (Nitecki and Nitecki, 1986). This early human predation probably focused on small or weak individuals rather than large, potentially dangerous animals. A major advance occurred in hunting proficiency 40,000 years ago, perhaps coinciding with the evolutionary emergence of modern people (Klein, 1984). For the first time, systematic and organized predation on large, powerful animals became an important part of human subsistence. It is likely that the first appreciable human effects on vertebrate faunas occurred at this time (Klein, 1992).

Nevertheless, only seven species of large mammals (one elephant, one horse, one camel, one deer, and three bovid species) disappeared from Africa during the late Pleistocene (Martin, 1984). This might be because African large mammals were hunted by humans while humans were evolving anatomically and behaviorally. As a result, most African mammals, to varying extents, were able to survive prehistoric human predation, because they had the opportunity to evolve in response to human predation as the humans evolved.

In the Americas, the first human occupation was by modern *Homo sapiens*, who was able to cross the Bering Strait land bridge because of lowered sea levels. The North American vertebrate fossil record of the past 2 million years reveals little extinction without replacement until the first humans arrived about 11,000 years ago (Martin, 1990). At that time, an overall warming trend was also causing major changes in the latitudinal, elevational, and edaphic ranges for most species of plants (Webb et al., 1993). The P-H extinction event primarily affected large-mammal species ($\geq$44 kg); nearly all amphibian, reptile, and small-mammal species survived. A variety of North American megafauna was lost, including two glyptodont, four ground sloth, two bear, two sabertooth, one cheetah, one beaver, two capybara, two mastodon, one mammoth, one horse, one tapir, two peccary, three camel, two deer, one pronghorn, and four bovid genera (Martin, 1984; 1990).

The collapse of the large North American mammal communities led to the demise of dependent species, such as carrion-feeding birds that fed on the abundant variety of carrion provided by herds of large animals, much as in modern African game parks. For example, 11,000 years ago, when ground sloths, mammoths, mastodons, horses, tapir, camels, and other species existed across North America, the currently endangered California condor lived as far away as Florida and New York (Steadman and Miller, 1987).

People spread rapidly across the Bering Strait and North America, and within only hundreds of years, descendants of these big-game hunters dispersed throughout Central and South America as well. Their spears tipped with stylistically distinctive projectile points have been found with the bones of extinct megafauna (especially mammoths) at sites radiocarbon-dated to within a century or two of 11,000 years ago.

According to the "Pleistocene overkill theory," early big-game hunters were the primary or only cause of the collapse of American large-mammal communities (Martin, 1984; 1990; Diamond, 1992). Scientists opposed to the overkill theory generally believe that changing climates and habitats were the sole or main cause of the P-H extinctions, but cannot explain why American large-mammal communities that survived many glacial-interglacial cycles quickly collapsed at the same time human populations in North America first became significant. Some scientists favor a combination of climate and habitat stress and human predation as the cause for the P-H extinctions (Owen-Smith, 1987). Regardless of the cause (Martin and Klein, 1984), the extinctions must have had a serious effect on North American ecosystems that once included a variety of grazers and browsers, not to mention their predators and commensals.

When climates warmed at the end of the last glaciation, plant communities of southern regions or low elevations did not simply move as entire communities. Instead, species of plants responded individually and, therefore, plant communities assembled during the Holocene were not the same as those that had existed farther south or at lower elevations during the late Pleistocene (Betancourt et al., 1990; Pielou, 1991). The distributions of surviving species of mammals also changed, again somewhat independently rather than as intact communities (Semken, 1983).

With most of the large mammals extinct, some peoples shifted to a more generalized diet, while others specialized in hunting bison, which survived into the Holocene. Bones from North American Holocene archaeological sites might represent as many as 50 species of vertebrates ranging from frogs and songbirds to eagles, bear, deer, and moose. Many of these sites bear evidence of species well outside their known post-Columbian range (Semken, 1983; Pielou, 1991). Intertribal trading of birds and mammals might account for some of these distributions, although many, if not most, reflect formerly indigenous populations, including species such as the trumpeter swan, Mississippi kite, swallow-tailed kite, whooping crane, sandhill crane, long-billed curlew, Carolina parakeet, ivory-billed woodpecker, common raven, fish crow, rice rat, Allegheny woodrat, fisher, and puma. Prehistoric hunting, trapping, habitat modification, and climate-driven habitat changes might have been involved in some of the contractions of species' ranges.

In parts of North America, the domesticated bottle gourd, chili pepper, beans, squash, and maize existed 5,000-6,000 years ago. The more sedentary lifestyle that came with agriculture was accompanied by a reduced dependence on wild plants and animals. But the gradual transition probably entailed shifting cultivation rather than long-term use of the same plots, and hunting, trapping, fishing, and gathering remained an important part of subsistence in most of North America long after plants and animals were domesticated. At the time of European contact, North American Indians still hunted turkey, black bear, deer, elk, moose, bison, and many smaller species. Some species, such as the California condor (Simons, 1983; Bates et al., 1993) and eagles, were hunted for feathers, bones, or ceremonial purposes rather than for food.

Agriculture affects ecosystems through the reduction or loss of certain species of plants and animals, the propagation of others, and modifications of landscape, soils, and water supply through deforestation, erosion, deposition, channeling, flooding, and draining (Steadman, 1995a). Use of pesticides, energy supplements, and grazing are other factors. Because of the good water supply and fertile alluvial soils, much early agriculture developed along major river valleys; consequently, riparian biotas have been disturbed for centuries, if not millennia (Rea, 1983).

Chaco Canyon, New Mexico, is a clear example of prehistoric overexploitation of natural resources, as revealed by studies of plant macrofossils from ancient packrat middens (Betancourt and Van Devender, 1981; Betancourt et al., 1986). The buildings constructed at Chaco Canyon from about AD 900 to 1150 by the Anasazi required 200,000 beams from large coniferous trees (ponderosa pine, subalpine fir, and blue/Englemann spruce). The trees were brought from as far away as 75 km, because these alpine conifers cannot live in the low elevation of Chaco Canyon. When the Anasazi arrived, the area around Chaco Canyon was a pinyon-juniper woodland. The pinyon and juniper trees provided firewood for the Anasazi, who farmed a narrow band of floodplain soil near their dwellings.

As time progressed, the Anasazi had to go 15 km or more from Chaco Canyon just to gather firewood. They had transformed the area from a pinyon-juniper woodland to a nearly woodless desert scrub. The removal of pinyon, juniper, and riparian trees produced soil erosion, increased runoff, and probably a lowered water table in Chaco Canyon, all of which made it difficult to continue irrigated farming of the valley bottom. Drought-stressed agriculture and lack of firewood might have caused the Anasazi to abandon Chaco Canyon about 700 years ago (Betancourt and Van Devender, 1981). Scattered junipers grow at Chaco Canyon today, but no pinyons do. The incised streambed would be virtually impossible to modify today for agriculture. Because of human activities that began more than 1,000 years ago, Chaco Canyon still sustains impoverished plant and animal communities and remains a poor place for people to live.

Another example of prehistoric humans having a significant effect on the carrying capacity of their ecosystems involves the Aleuts, whose overexploitation of sea otters appears to have changed the coastal ecosystems of the Aleutian Islands as long as 2,500 years ago (Simenstad et al., 1978).

PREHISTORIC HUMAN IMPACT ON ISLAND ECOSYSTEMS

The relatively small land areas of islands result in small populations of organisms that tend to be more vulnerable to extinction than those on continents (MacArthur and Wilson, 1967). Compared with continental species, a much greater proportion of island species have become extinct or endangered during the past several centuries of exploration and exploitation. In the past 2 decades, we have learned that, at least for vertebrates, human-caused extinctions had occurred on the world's islands during prehistoric times. Those extinctions seem to have been due to the same processes that lead to the extinction of island species today: direct human predation, predation from introduced mammals (such as rats, dogs, and pigs), habitat changes (such as those caused by deforestation, agriculture, and introduction of exotic plants), and introduced pathogens (such as avian malaria and avian pox).

One way to assess the environmental impacts of prehistoric peoples on oceanic islands is to study the biotic history of islands that never were inhabited in pre-Columbian times. According to biogeographic theory, background extinctions are a regular part of an island's biological heritage. Knowing the rate of background extinction allows the severity of the human-caused extinction we see today to be evaluated. The human history of the Galápagos Islands, for example, began with brief Spanish and British visits in the 16th and 17th centuries. We can examine the pre-Columbian Galápagos biota to determine the level of background extinction.

Thirty-four species of reptiles, birds, and mammals are known to have become extirpated in the Galápagos during the Holocene (Steadman et al.,

1991). The Holocene vertebrate fossil record from the Galápagos is based upon 15 sites from five different islands, with nearly 500,000 identified bones spanning the past 8,000 years. This record reveals only three extinct populations or species that might have become extinct before human contact. In other words, 31 of the 34 known extinctions occurred after the arrival of people 200 years ago. A similar pattern has been corroborated in Tonga, where the arrival of humans 3,000 years ago caused more extinctions of birds than had occurred during the previous 100,000 years (Steadman, 1993).

Approximately 9,600 species of birds exist today (Sibley and Monroe, 1990). The prehistoric human colonization of Pacific islands (Melanesia, Micronesia, and Polynesia) resulted in the loss of as many as 2,000 species of birds, especially flightless rails, but also petrels, ibises, herons, ducks and geese, hawks and eagles, megapodes, sandpipers, gulls, pigeons and doves, parrots, owls, and passerines (Steadman, 1989; 1991; 1993; 1995b). The world avifauna would be about 20% richer today had islands of the Pacific remained unoccupied by humans.

In the Hawaiian Islands, at least 77 endemic species of birds have become extinct since the arrival of Polynesians nearly 2,000 years ago (James and Olson, 1991; Olson and James, 1991) (see Table 2-1). Furthermore, several individual island populations of species have been reduced or extirpated. Most extinct species of Hawaiian birds, including the majority of

TABLE 2-1 Loss of Birds in the Hawaiian Islands Since Human Arrival[a]

Species	Number of Extinct Species	Number of Extirpated Populations
Petrels	1	2
Ibises	2	3
Waterfowl	9	17
Rails	12	14
Hawks	2	6
Owls	4	4
Crows	2	5
Warblers	0	1
Thrushes	1	4
Honeyeaters	6	9
Drepanidine finches	38	56
Total	77	121

[a]Data condensed from James and Olson (1991) and Olson and James (1991).

cardueline finches (the "Hawaiian honeycreepers"), died out before the arrival of Europeans (Olson and James, 1982; 1984; James et al., 1987).

On hundreds of tropical islands, including those that are a part of or affiliated with the United States (Puerto Rico, U.S. Virgin Islands, Hawaii, American Samoa, and various Micronesian island groups), the prehistoric and ongoing extinction of numerous populations and species of birds has consequences beyond losing the birds themselves. Most of the extinct land birds were omnivores, frugivores, granivores, or nectarivores; indigenous plants might have depended on those birds for pollination or seed dispersal. Many species of Pacific island trees and shrubs seem to have no natural means of intra- or interisland dispersal today. Because island biotas are so degraded (i.e., so many populations and species already are extinct), endangered-species programs on islands face an extraordinary challenge to maintain conditions that might permit the long-term survival of the remaining species.

RELATING THE PAST TO THE PRESENT

Ecosystem Degradation and Restoration

Environmental problems often are viewed as modern phenomena. It is true that most plant and animal communities have been affected by the past 2 centuries of commercial and residential development; nonetheless, most of the earth was far from pristine in preindustrial times. The use of tools and fire has set humans apart from other animals for many millennia, and all ancient human societies had various effects on the ecosystems within which they lived (e.g., Simenstad et al., 1978; Betancourt and Van Devender, 1981; Betancourt et al., 1986; Diamond, 1992).

By studying the effects prehistoric peoples had on the environment, we can begin to estimate the composition of plant and animal communities during two critical times: before human contact (when communities were pristine) and at western contact (when all previous human effects had been due to native peoples). Such knowledge is important for long-term projections of community stability, as well as to help understand to what extent various communities might recover from human disturbances.

This is not to suggest that the environmental conditions that prevailed in 1492, when major European exploration of the Americas began, be the benchmark for discussion of the condition of any given ecosystem or a goal for restoration. First, North America was not pristine then. The last time our continent was unaffected by human activity was late in the last glacial interval, when a dramatically different climate resulted in plant and animal associations that simply would not be possible in today's climate. In addition, exact reconstructions of 1492 biotic communities would be impossible because so many species, subspecies, and varieties of plants and animals already have been lost. The grizzly bear, gray wolf, and mountain lion, for example, no longer occupy more than a million km^2 of their former North American ranges (Mech, 1974; Currier, 1983; Pasitschniak-Arts, 1993). Given current habitat conditions and land-use patterns, restoration of these species would be impractical in much of their former ranges.

However, we still can use information about the prehistoric environment data to help plan for the future. Many tracts of land in North America are relatively undisturbed, and if biologically feasible, we might try to restore locally extirpated species as part of a recovery program for endangered species. Rather than using past distributions as strict guidelines for restoration efforts, the data could be part of planning conservation programs, with a goal of preserving plant and animal communities that approach those of a less disturbed state. In some cases, such as on Pacific islands, prehistoric information provides clear direction for translocating endangered species onto islands that once were part of their natural range (Franklin and Steadman, 1991).

Rates of Extinction

In a 1992 report to Congress, the U.S. Fish and Wildlife Service categorized 711 species of plants and animals as follows: improving—69 (10%), stable—201 (28%), declining—232 (33%), extinct—14 (2%), and unknown—195 (27%) (USFWS, 1992). The relative proportions of these categories among plants and animals are similar, except for extinctions, all of which are animals. The 232 declining species eventually will become extinct unless the decline is halted.

Rates of extinction are difficult to quantify precisely, in part because various authors measure them in different ways (Benton, 1995). However, examinations of relative rates of extinctions and of well-known biota make it clear that extinctions are increasing in many groups of organisms (Nott et al., 1995).

For example, Nott et al. provided estimates of regional and global extinction rates based on some well-known examples. They listed 36 species that had become extinct in the past 100 years of the 8,500 species in the South African floristic community known as *fynbos*; another 618 were deemed to be at some risk of extinction. Forty taxa—mostly species—of the approximately 950 species of freshwater fishes in the United States became extinct in the past 100 years, with more than 100 at risk of extinction. Of the 297 North American species and 13 subspecies of freshwater mussels in North America, 21 have become extinct since the beginning of this century and 120 are endangered. Sixty species of mammals have become extinct in the recent past; of those, 18 were found among the 300 species in the nonmarine Australian mammal fauna. An additional 43 species have been lost from more than 50% of their former ranges. (Other well-known cases, such as birds and plants on islands and the cichlid fishes of Lake Victoria, are discussed elsewhere in this chapter.)

Nott et al. (1995) made a conservative estimate of global extinction rates for various taxonomic groups by assuming that no other species in those groups had become extinct worldwide and then dividing the number of regional extinctions by the global number of species in the groups. The resulting estimated global extinction rates—which would have been larger if other known extinctions were included in the estimate—ranged from 10 to 1,000 times the estimated background rates. They predicted that future extinction rates, based on knowledge of species currently at risk of extinction, would be even higher.

Part of the difficulty in grasping the significance of extinction rates is that they should be evaluated in intervals of millennia or more, as well as the seasons, years, or decades that measure most ecological studies. Another well-studied taxonomic group is North American birds. At least 20 to 40 species of birds were lost at the end of the Pleistocene, mainly because of the extinction of large mammals on which they depended (Steadman and Martin, 1984; Steadman and Miller, 1987; Emslie 1987, 1990). In the next 10,000 years, before the arrival of Europeans, only two North American species definitely became extinct, the flightless marine duck *Chendytes lawi* (Morejohn, 1976; Guthrie, 1992) and the small turkey *Meleagris crassipes* (Rea, 1980).

In the past 200 years, at least five species of birds have been lost (great auk, Labrador duck, passenger pigeon, Carolina parakeet, and ivory-billed woodpecker). Two others, the Eskimo curlew and Bachman's warbler, are near extinction. Several other species persist today in small, local populations (e.g., California condor, whooping crane, red-cockaded woodpecker, black-capped vireo, golden-cheeked warbler, and Kirtland's warbler). Some of those last eight species are likely to die out in the next 200 years, as are others that are known to be endangered, threatened, or in decline today (see North American Breeding Bird Survey data for examples of birds whose populations have declined substantially in recent years).

Most climatologists believe that the world will experience another ice age "unless there is some fundamental and unforeseen change in the climate system" (Imbrie and Imbrie, 1979). Indeed, the earth's climate has varied as far back as we can tell anything about it and almost surely will continue to vary. If the extinction rate of birds continues to be about five species per 200 years or about 25 species per millennium, there could be hundreds fewer species to face the changing habitats that will come with North America's next ice age or any major climatic change. Speciation (i.e., the evolution of new species) is unlikely to offset these losses. The extinction of birds could affect other organisms as well. For example, many species of North American plants may face a northward range shift if the climate warms (or a southward shift if it cools); they will have to accomplish this task without the passenger pigeon, which was by far the most abundant consumer and disperser of their seeds until a century ago.

Habitat Loss

Species endangerment and extinction have three major anthropogenic causes—overhunting or overharvesting; introduction of nonnative species, including the spread of disease; and habitat degradation or loss. All three causes probably were factors in prehistoric as well as modern times. In some locations and for certain species, the first two have been the most critical. For example, the decline of some species of marine vertebrates, such as certain whales, can be attributed largely to commercial overexploitation. And many birds in the Hawaiian Islands have suffered from predation and diseases caused by introduced species (Olson and James, 1982; Ehrlich et al., 1992).

For most species in decline and for most of those on the edge of extinction in the U.S. today, however, the most serious threat appears to be habitat degradation or loss (hereafter denoted as habitat destruction). Habitat destruction is the primary threat to the majority of endangered and threatened plants (Cook and Dixon, 1989) and is an increasing threat to songbirds and freshwater fishes (Miller et al., 1989; Ehrlich et al., 1992; Allen and Flecker, 1993; Noss et al., 1995) and perhaps for hundreds or even thousands of the lesser-known invertebrates in the United States (Deyrup and Franz, 1994).

Habitat destruction is described either as the current rate of destruction (expressed in hectares/year) or as a cumulative amount of destruction (expressed as a percentage of some historic baseline). No comprehensive summary has been compiled of the current rates of destruction of major habitat types, but several studies have reviewed cumulative losses in particular regions, such as California (Jensen et al., 1993), and for particular habitat types such as wetlands (NRC, 1992a; 1995). A recent review of cumulative loss by Noss et al. (1995) is noteworthy for its broad scope and fine level of distinction with which it treats habitat types and locations. Noss et al. acknowledged that the information needed to provide uniform geographic coverage is missing, and that habitat destruction is better catalogued in the eastern United States than elsewhere.

The discussion of cumulative losses that follows should be read with three considerations. First, habitat that is degraded or lost is not necessarily biologically depauperate. For example, grazed western grasslands and shrub steppe still support wildlife,[1] just not the diversity that inhabited those lands before grazing. Second, even habitat that is not considered degraded or lost might not be able to support the abundance and variety of species that it once did. Such habitat might be used for recreation or other activities, or simply little of it might be left. Moreover, some habitats are fragmented or of such small area that they might no longer include the habitats required by some specialized species or might be below the critical area needed to support species with large home ranges. Finally, air or water pollution and climate change might be affecting the viability of habitat too subtly for the consequences to be revealed in surveys.

Cumulative losses of wetlands have attracted much concern and scrutiny (NRC, 1992a), perhaps because 50% of animals and 33% of plants listed as endangered or threatened under the Endangered Species Act depend on wetland habitats (Nelson, 1989). Although wetlands comprise only a small percentage of the U.S. land area, they have been prime sites for conversion to agriculture or urban sprawl. According to the National Research Council (NRC, 1992a), approximately 30% (117 million acres) of U.S. wetlands have been converted (53%, if Alaska is excluded). In some regions and for some types of wetlands, the losses have been more severe. For example, California has lost at least 80% of its interior and coastal wetlands (Jensen et al., 1993). More than 85% of the flow of inland waters is now artificially controlled, and 98% of stream reaches has been degraded enough to be unworthy of federal designation as wild or scenic rivers (Benke, 1990).

The cumulative losses of terrestrial and aquatic habitat reviewed by Noss et al. (1995) included 27 specific habitats that they classified as "critically endangered" (greater than 98% cumulative destruction). Those habitats included longleaf pine forests in the southeastern coastal plain, tallgrass prairie east of the Missouri River, dry prairie in Florida, native grasslands in California, and old-growth pine forests in Michigan. Other habitats, classified as endangered (between 85% and 98% cumulative destruction) included old-growth forest in the Pacific Northwest and elsewhere, red spruce forests in the central Appalachians, coastal heathland in southern New England and Long Island, all other tall-grass prairie, vernal pools in the Central Valley and Southern California, coastal redwood forests in California, and native shrub and grassland steppe in Oregon and Washington.

Habitat loss is a direct threat to many species, and therefore important to administration of the Endangered Species Act. Indeed, the ESA's purpose "to provide a means whereby the ecosystems upon which endangered

---

[1]Western grasslands are considered by some to be "healthier" than they were 100 years ago, although information is not available to assess this objectively (NRC, 1994).

species and threatened species depend may be conserved" recognizes the importance of habitat to species, and for this reason, we return to the subject of habitat protection under the ESA in Chapters 4 and 5, where we describe modern ecological concepts of habitat, clarify the role of habitat in the protection of endangered species, and suggest ways in which the act's stated purpose can be better achieved.

Introduced Species

Much of the world's biota has been modified by the introduction of exotic (nonnative) species. Introduced plant species range from an estimated 7% of the total in Java to 28% in Canada and 47% in New Zealand (Heywood, 1989); introduced mammal species constitute about 95% of the total in New Zealand and Hawaii but much less than 5% in North America (Brown, 1989). Introduced birds are nearly 70% of the total in Hawaii (Brown, 1989), with "virtually no native land birds below 1000 m" (Pimm, 1989). Fish are among the most widely introduced of vertebrates; in many cases, their introductions were deliberate, as vividly described for the western United States by Lampman (1946). Even in continental areas, exotic fishes can constitute large fractions of the total number of species and have deleterious effects (Courtenay and Stauffer, 1984). For example, in California, 47 species of a total of 112 freshwater and anadromous species are introduced (Moyle, 1976); introduced species constitute substantial portions of the biota in other western states as well (e.g., Sigler and Sigler, 1987; Sublette et al., 1990).

Despite the difficulty of proving that introduced species have caused extinctions, the circumstantial evidence is often overwhelming. Many extinctions, especially of birds and plants but also mammals, reptiles, snails, and others have been attributed to introduced species, especially on islands (Loope and Mueller-Dombois, 1989; Macdonald et al., 1989; Pimm, 1989) and in their analogs, isolated freshwater bodies (Ono et al., 1983). Groombridge (1992) considered that introduced species were responsible for 39% of all animal extinctions whose causes were known, and Macdonald et al. concluded that 12.7% of threatened terrestrial vertebrate species on mainland areas and 31% on islands were affected by introductions. For example, the National Research Council attributed the near extinction of the endangered Hawaiian crow or 'Alala (*Corvus hawaiiensis*), to introduced predators, introduced diseases, and habitat alteration (NRC, 1992b). The European zebra mussel (*Dreissena polymorpha*) (Ricciardi et al., in press) and the Asiatic clam (*Corbicula fluminea*) (Clarke, 1988) have been implicated in the extinctions of species of native freshwater mussels in the United States, especially in drainages of the Gulf of Mexico and Atlantic Ocean; those authors consider that further extinctions are likely. Solz and Naiman (1978) and Ono et al. (1983) attributed the extinction of the Ash Meadows (Nevada) killifish (*Empetrichthys merriami*) to competition and predation from introduced species, and Skelton (1993), Etnier and Starnes (1993), and Jenkins and Burkhead (1994) described local extirpation or near extinction of rare fish taxa by exotics and habitat degradation in South Africa, Tennessee, and Virginia. In the western United States and elsewhere, introduced salmonid species have hybridized with endemic forms (Behnke, 1992). Perhaps the most notorious of all exotic fish introductions is that of the Nile perch (*Lates nilotica*) into Lake Victoria in East Africa, where it has devastated the diversity of the endemic cichlid fishes. Estimates of extinctions there have been complicated by habitat changes and incomplete sampling, but Witte et al. (1992) estimated that 200 species of endemic cichlid fishes (of an original total of about 300) had "already disappeared or [were] threatened with extinction" mainly as a result of predation by the Nile perch.

In many cases, the effects of exotic species have been exacerbated by habitat degradation. For example, loss of habitat reduced the Hawaiian crow population and made it more vulnerable to the effects of introduced predators and diseases (NRC, 1992b). Impoundments on the Colorado River changed the habitat to favor introduced species of fishes that compete with and prey on some native endangered species in addition to reducing spawning habitat for the native species and adversely changing the water temperature (Ono et al., 1983).

Accidental species introductions have increased with increased human mobility and are not likely to decline in the near future. Deliberate introductions have occurred for hundreds of years and continue despite a growing awareness of the problems they can cause and laws and regulations that prohibit them. In a few cases, introductions are extremely well documented and studied as, for example, the various predatory species introduced into California to control the California red scale (*Aonidiella aurantii*), which was itself accidentally introduced from Australia between 1868 and 1875 (Luck, 1986). In general, the effects of introduced species are difficult to predict (NRC, 1986), although some progress has been made (Levin, 1989; Pimm, 1989; Simberloff, 1989; Ricciardi et al., in press). Unless they occupy small isolated areas, such as small islands or lakes, exotics are difficult to eradicate, although control is sometimes possible if started early, especially for many vertebrates (Groves, 1989; Macdonald et al., 1989).

CONCLUSIONS AND RECOMMENDATIONS

Several concepts from this chapter apply to the understanding of biological diversity and attempts to protect it. From these concepts we draw the following conclusions.

• The current extinction event differs from extinction events in the fossil record in being less selective, i.e., it actually or potentially involves all taxonomic groups of organisms, in any sort of nonmarine habitat. Although the number of documented extinctions in recent years might appear to be small as a fraction of all extant species, it is important to understand that even so-called catastrophic extinctions took many thousands of years to occur. Thus, the appropriate focus is the *rate* of extinctions, and the current extinction rate appears to be significantly greater than background rates. Available evidence suggests that the current accelerated extinction rate is largely human-caused and is likely to increase rather than decrease in the near future.

• Both modern and prehistoric human activities, especially those associated with agriculture and urbanization, have altered natural ecosystems such as forests, lakes, prairies, and floodplains through the reduction or loss of certain species of plants and animals, the propagation of others, and modifications of landscape, soils, and water supply.

• On islands, including those that are a part of or affiliated with the United States, a relatively high percentage of species has become extinct or endangered because of both modern and prehistoric human activities. Direct human predation, predation from introduced mammals (such as rats, dogs, and pigs), habitat changes (such as deforestation, urbanization, agriculture, and exotic plants), and introduced pathogens (such as avian pox and avian malaria) account for these extinctions. Because island biotas already have lost so many populations and species and land resources are so limited, conservation programs on islands face an extraordinary challenge to maintain conditions that will allow the long-term survival of existing species.

• The prehistoric and ongoing extinction of so many populations and species of birds on islands has consequences beyond losing the birds themselves. The losses of land birds, for example, may influence pollination or seed dispersal of indigenous plants.

Based on these conclusions and as substantiated within this chapter, we make these recommendations.

• Because we can relate many biotic events to abiotic events in earth history with reasonable accuracy, our knowledge of the past is important for predicting biotic responses to future changes in climate and other physical phenomena. Therefore, when feasible, conservation decisions should consider long-term impacts as revealed by our improved understanding of the earth's physical and biotic history.

• The natural and anthropogenic processes affecting extinctions and speciation operate at various time scales. To understand the potential long-term impacts of the current extinction event, the process of extinction should be examined across these time scales. Extinction rates that seem low on first inspection could in fact result in major losses of species if evaluated on longer time scales.

• There are three primary anthropogenic processes that lead to species endangerment and extinction—overharvesting; the introduction of nonnative species, including the spread of disease; and habitat destruction. For most endangered species in the United States today, the most serious threat is habitat destruction. Because of this, habitat conservation is the best single means to counter extinction. Introduced species also present a significant threat. Policies that prevent deliberate introductions and reduce the adverse effects of accidental introductions should be encouraged.

REFERENCES

Allen, J.D., and A.S. Flecker. 1993. Biodiversity conservation in running waters. BioScience 43:32-43.

Appenzeller, T. 1993. Ancient climate coolings are on thin ice. Science 262:1818-1819.

Bates, C.D., J.A. Hamber, and M.J. Lee. 1993. The California condor and the California Indians. Am. Indian Art 19:40-47.

Behnke, R.J. 1992. Native Trout of Western North America. American Fisheries Society, Bethesda, Md.

Benke, A.C. 1990. A perspective on America's vanishing streams. J. Am. Benthol. Soc. 9(1):77-78.

Benton, M.J. 1995. Diversification and extinction in the history of life. Science 268:52-58.

Betancourt, J.L., and T.R. Van Devender. 1981. Holocene vegetation in Chaco Canyon, New Mexico. Science 214:656-658.

Betancourt, J.L., J.S. Dean, and H.N. Hull. 1986. Prehistoric long-distance transport of construction beams, Chaco Canyon, New Mexico. Am. Antiquity 51:370-375.

Betancourt, J.L., T.R. Van Devender, and P.S. Martin, eds. 1990. Packrat Middens: The Last 40,000 Years of Biotic Change. Tucson, Ariz.: University of Arizona Press.

Brown, J.H. 1989. Patterns, modes, and extents of invasions by vertebrates. Pp. 85-109 in Biological Invasions: A Global Perspective. SCOPE 37, J.A. Drake, H.A. Mooney, F. di Castri, R.H. Groves, F.J. Kruger, M. Rejmánek, and M. Williamson, eds. Chichester, U.K.: John Wiley & Sons.

Burney, D.A. 1993. Recent animal extinctions: Recipes for disaster. Am. Sci. 81:530-541.

Clarke, A.H. 1988. Aspects of corbiculid-unionid sympatry in the United States. Malacol. Data Net 2(3/4):57-99.

Cook, R.E., and P. Dixon. 1989. A review of recovery plans for threatened and endangered plant species. World Wildlife Fund Report, Washington, D.C.

Courtenay, W.R., Jr., and J.R. Stauffer, Jr. 1984. Distribution, Biology, and Management of Exotic Fishes. Baltimore, Md.: The Johns Hopkins University Press.

Currier, M.J.P. 1983. *Felis concolor*. Mamm. Species 20.

Dansgaard, W., S.J. Johnsen, H.B. Clausen, D. Dahl-Jensen, N.S. Gundestrup, C.U. Hammer, C.S. Hvidberg, J.P. Steffensen, A.E. Sveinbjörnsdottir, J. Jouzel, and G. Bond. 1993. Evidence for general instability of past climate from a 250-kyr ice-core record. Nature 364:218-220.

Deyrup, M., and R. Franz, eds. 1994. Rare and Endangered Biota of Florida. Invertebrates, Vol. 4. Gainesville, Fla.: University Presses of Florida.

Diamond, J. 1992. The Third Chimpanzee. New York: Harper Collins.

Ehrlich, P.R., D.S. Dobkin, and D. Wheye. 1992. Birds in Jeopardy. Stanford, Calif.: Stanford University Press.

Emslie, S.D. 1987. Age and diet of fossil California condors in Grand Canyon, Arizona. Science 237:768-770.

Emslie, S D. 1990. Additional ^{14}C dates on fossil California condor. Natl. Geograph. Res. 6:134-135.

Engbring, J. and H.D. Pratt. 1985. Endangered birds in Micronesia: Their history, status, and future prospects. Pp. 71-105 in Bird Conservation 2, S.A. Temple, ed. Madison, Wisc.: University of Wisconsin Press.

Etnier, D.A., and W.C. Starnes. 1993. The Fishes of Tennessee. Knoxville, Tenn.: University of Tennessee Press.

Franklin, J., and D.W. Steadman. 1991. The potential for conservation of Polynesian birds through habitat mapping and species translocation. Conserv. Biol. 5:506-521.

FWS (U.S. Fish and Wildlife Service). 1992. Endangered and Threatened Species Recovery Program. Report to Congress. U.S. Department of Interior, U.S. Fish and Wildlife Service, Washington, D.C.

GRIP (Greenland Ice-core Project). 1993. Climate instability during the last interglacial period recorded in the GRIP ice core. Nature 364:203-207.

Groombridge, B., ed. 1992. Global Biodiversity: Status of the Earth's Living Resources. New York: Chapman and Hall.

Grootes, P.M., M. Stuiver, J.W.C. White, S. Johnsen, and J. Jouzel. 1993. Comparison of oxygen isotope records from the GISP2 and GRIP Greenland ice cores. Nature 366:552-554.

Groves, R.H. 1989. Ecological control of invasive terrestrial plants. Pp. 437-461 in Biological Invasions: A Global Perspective. SCOPE 37, J.A. Drake, H.A. Mooney, F. Di Castri, R.H. Groves, F.J. Kruger, M. Rejmanek, and M. Williamson, eds. Chichester, U.K.: John Wiley & Sons.

Guthrie, D.A. 1992. A late Pleistocene avifauna from San Miguel Island, California. Sci. Ser. Publ. 36, pp. 319-327. Natural History Museum of Los Angeles County, Los Angeles, Calif.

Heywood, V.H. 1989. Extents and modes of terrestrial invasions by plants. Pp. 31-60 in Biological Invasions: A Global Perspective. SCOPE 37, J.A. Drake, H.A. Mooney, F. Di Castri, R.H. Groves, F.J. Kruger, M. Rejmanek, and M. Williamson, eds. Chichester, U.K.: John Wiley & Sons.

Imbrie, J., and K.P. Imbrie. 1979. The Ice Ages: Solving the Mystery. Hillside, N.J.: Enslow Publishers.

Jablonski, D. 1991. Extinctions: A paleontological perspective. Science 253:754-757.

James, H.F., and S.L. Olson. 1991. Descriptions of thirty-two species of birds from the Hawaiian Islands. Part II. Passeriformes. Ornithol. Monogr. 46:1-88.

James, H.F., T.W. Stafford, Jr., D.W. Steadman, S.L. Olson, P.S. Martin, A.J.T. Jull, and P.C. McCoy. 1987. Radiocarbon dates on bones of extinct birds from Hawaii. Proc. Natl. Acad. Sci. USA 84:2350-2354.

Jenkins, R.E., and N.M. Burkhead. 1994. Freshwater Fishes of Virginia. American Fisheries Society, Bethesda, Md.

Kerr, R.A. 1993a. The whole world had a case of the Ice Age shivers. Science 262:1972-1973.

Kerr, R.A. 1993b. A bigger death knell for the dinosaurs? Science 261:1518-1519.

Klein, R.G. 1984. Mammalian extinctions and Stone Age people in Africa. Pp. 553-573 in

Quaternary Extinctions, P.S. Martin and R.G. Klein, eds. Tucson, Ariz.: University of Arizona Press.

Klein, R.G. 1992. The impact of early people on the environment: The case of large mammal extinctions. Pp. 13-34 in Human Impact on the Environment: Ancient Roots, Current Challenges, J.E. Jacobsen and J. Firor, eds. Boulder, Colo.: Wheatview.

Klein, R.G. 1994. The problem of modern human origins. Pp. 3-17 in Origins of Anatomically Modern Humans, M.H. Nitecki and D.V. Nitecki, eds. New York: Plenum.

Lampman, B.H. 1946. The Coming of the Pond Fishes. Portland, Oreg.: Binfords & Mort.

Levin, S.A. 1989. Analysis of risk and invasions and control programs. Pp. 425-435 in Biological Invasions: A Global Perspective. SCOPE 37, J.A. Drake, H.A. Mooney, F. Di Castri, R.H. Groves, F.J. Kruger, M. Rejmanek, and M. Williamson, eds. Chichester, U.K.: John Wiley & Sons.

Loope, L.L., and D. Mueller-Dombois. 1989. Characteristics of invaded islands, with special reference to Hawaii. Pp. 257-280 in Biological Invasions: A Global Perspective. SCOPE 37, J.A. Drake, H.A. Mooney, F. Di Castri, R.H. Groves, F.J. Kruger, M. Rejmanek, and M. Williamson, eds. Chichester, U.K.: John Wiley & Sons.

Luck, R.F. 1986. Biological control of California red scale. Pp. 165-189 in Ecological Knowledge and Environmental Problem-Solving: Concepts and Case Studies. Washington, D.C.: National Academy Press.

MacArthur, R.H., and E.O. Wilson. 1967. The Theory of Island Biogeography. Princeton, N.J.: Princeton University Press.

Macdonald, I.A.W., L.L. Loope, M.B. Usher, and O. Hamman. 1989. Wildlife conservation and the invasion of nature reserves by introduced species: A global perspective. Pp. 215-255 in Biological Invasions: A Global Perspective. SCOPE 37, J.A. Drake, H.A. Mooney, F. Di Castri, R.H. Groves, F.J. Kruger, M. Rejmanek, and M. Williamson, eds. Chichester, U.K.: John Wiley & Sons.

Markgraf, V. 1989. Palaeoclimate in central and South America since 18,000 BP based on pollen and lake-level records. Quat. Sci. Rev. 8:1-24.

Martin, P.S. 1984. Prehistoric overkill: The global model. Pp. 354-403 in Quaternary Extinctions, P.S. Martin and R.G. Klein, eds. Tucson, Ariz.: University of Arizona Press.

Martin, P.S. 1990. 40,000 years of extinctions of the "planet of doom." Palaeogeogr. Palaeoclimatol. Palaeoecol. 82:187-201.

Martin, P.S., and R.G. Klein, eds. 1984. Quaternary Extinctions. Tucson, Ariz.: University of Arizona Press.

Martinson, D.G., N.G. Pisias, J.D. Hays, J. Imbrie, T.C. Moore, Jr., and N.J. Shackleton. 1987. Age dating and the orbital theory of the Ice Ages: Development of a high-resolution 0 to 300,000-year chronostratigraphy. Quat. Res. 27:1-29.

Mayewski, P.A., L.D. Meeker, S. Whitlow, M.S. Twickler, M.C. Morrison, R.B. Alley, P. Bloomfield, and K. Taylor. 1993. The atmosphere during the Younger Dryas. Science 261:195-197.

Mech, L D. 1974. *Canis lupus*. Mamm. Species 37.

Miller, R.R., J.D. Williams, and J.E. Williams. 1989. Extinctions of North American fishes during the past century. Fisheries 14:22-38.

Morejohn, G.V. 1976. Evidence of the survival to recent times of the extinct flightless duck *Chendytes lawi* Miller. Smithsonian Contrib. Paleobiol. 27:207-211.

Morell, V. 1993. How lethal was the K-T impact? Science 261:1518-1519.

Moyle, P.B. 1976. Inland Fishes of California. Berkeley, Calif.: University of California Press.

Nadachowski, A. 1993. The species concept and Quaternary mammals. Quat. Int. 19:9-11.

Nelson, J. 1989. Agriculture, wetlands, and endangered species: The Food Security Act of 1985. Endangered Species Tech. Bull. 14:6-8.

Nitecki, M.H., and D.V. Nitecki, eds. 1986. The Evolution of Human Hunting. New York: Plenum.

Noss, R.F., E.T. LaRoe, III, and J.M. Scott. 1995. Endangered Ecosystems of the United States: A Preliminary Assessment of Loss and Degradation. Biol. Rep. 28. U.S. Department of the Interior, National Biological Service, Washington, D.C.

Nott, M.P., E. Rogers, and S. Pimm. 1995. Modern extinctions in the kilo-death range. Curr. Biol. 5:14-17.

Nowak, R.M., and J.L. Paradiso. 1983. Walker's Mammals of the World. Baltimore, Md.: The Johns Hopkins University Press.

NRC (National Research Council). 1986. Ecological Knowledge and Environmental Problem-Solving: Concepts and Case Studies. Washington, D.C.: National Academy Press.

NRC (National Research Council). 1992a. Restoration of Aquatic Ecosystems: Science, Technology, and Public Policy. Washington, D.C.: National Academy Press.

NRC (National Research Council). 1992b. The Scientific Bases for the Preservation of the Hawaiian Crow. Washington, D.C.: National Academy Press.

NRC (National Research Council). 1994. Rangeland Health: New Methods to Classify, Inventory, and Monitor Rangelands. Washington, D.C.: National Academy Press.

NRC (National Research Council). 1995. Wetlands: Characteristics and Boundaries. Washington, D.C.: National Academy Press.

Olson, S.L., and H.F. James. 1982. Fossil birds from the Hawaiian Islands: Evidence for wholesale extinction by man before western contact. Science 217:633-635.

Olson, S.L., and H.F. James. 1984. The role of Polynesians in the extinction of the avifauna of the Hawaiian Islands. Pp. 768-780 in Quaternary Extinctions, P.S. Martin and R.G. Klein, eds. Tucson, Ariz.: University of Arizona Press.

Olson, S.L., and H.F. James. 1991. Descriptions of thirty-two species of birds from the Hawaiian Islands. Part I. Non-passeriformes. Ornithol. Monogr. 45:1-88.

Ono, R.D., J.D. Williams, and A. Wagner. 1983. Vanishing Fishes of North America. Washington, D.C.: Stone Wall.

Owen-Smith, N. 1987. Pleistocene extinctions: The pivotal role of megaherbivores. Paleobiology 13:351-362.

Pasitschniak-Arts, M. 1993. Ursus arctos. Mamm. Species 439.

Pielou, E.C. 1991. After the Ice Age. Chicago: University of Chicago Press.

Pimm, S.L. 1989. Theories of predicting success and impact of introduced species. Pp. 351-367 in Biological Invasions: A Global Perspective. SCOPE 37, J.A. Drake, H.A. Mooney, F. Di Castri, R.H. Groves, F.J. Kruger, M. Rejmanek, and M. Williamson, eds. Chichester, U.K.: John Wiley & Sons.

Pirazzoli, P.A. 1993. Global sea-level changes and their measurement. Glob. Planet. Change 8:135-148.

Plummer, L.N. 1993. Stable isotope enrichment in paleowaters of the southeast Atlantic coastal plain, United States. Science 262:2016-2020.

Rampino, M.R., and S. Self. 1992. Volcanic winter and accelerated glaciation following the Toba super-eruption. Nature 359:50-52.

Raup, D.M., and D. Jablonski. 1993. Geography of end-Cretaceous marine bivalve extinctions. Science 260:971-973.

Rea, A.M. 1980. Late Pleistocene and Holocene turkeys in the Southwest. Natural History Museum of Los Angeles County Contrib. Sci. 330:209-224.

Rea, A.M. 1983. Once a River. Tucson, Ariz.: University of Arizona Press.

Retallack, G.J. 1995. Permian-Triassic life crisis on land. Science 267:77-80.

Ricciardi, A., F.G. Whoriskey, and J.B. Rasmussen. In press. Predicting the density and impact of *Dreissena* infestation on native unionid bivalves from *Dreissena* field density. Can. J. Fish. Aquat. Sci.

Schorger, A.W. 1955. The Passenger Pigeon: Its Natural History and Extinction. Madison, Wisc.: University of Wisconsin Press.

Semken, H. A., Jr. 1983. Holocene mammalian biogeography and climatic change in the eastern and central United States. Pp. 182-207 in Late-Quaternary Environments of the United States, Vol. 2, H.E. Wright, ed. Minneapolis, Minn.: University of Minnesota Press.

Shackleton, N.J. 1987. Oxygen isotopes, ice volume and sea level. Quat. Sci. Rev. 6:183-190.

Sharpton, V.L., K. Burke, A. Camargo-Zanoguera, S.A. Hall, D.S. Lee, L.E. Marín, G. Suárez-Reynoso, J.M. Quezada-Muñeton, P.D. Spudis, and J. Urritia-Fucugauchi. 1993. Chicxulub multiring impact basin: Size and other characteristics derived from gravity analysis. Science 261:1564-1567.

Sibley, C.G., and B.L. Monroe, Jr. 1990. Distribution and Taxonomy of Birds of the World. New Haven, Conn.: Yale University Press.

Sigler, W.F., and J.W. Sigler. 1987. Fishes of the Great Basin: A Natural History. Reno, Nev.: University of Nevada Press.

Simberloff, D. 1989. Which insect introductions succeed and which introductions fail? Pp. 61-75 in Biological Invasions: A Global Perspective. SCOPE 37, J.A. Drake, H.A. Mooney, F. Di Castri, R.H. Groves, F.J. Kruger, M. Rejmanek, and M. Williamson, eds. Chichester, U.K.: John Wiley & Sons.

Simenstad, C.A., J.A. Estes, and K.W. Kenyon. 1978. Aleuts, sea otters, and alternate stable-state communities. Science 200:403-411.

Simons, D.D. 1983. Interactions between California condors and humans in prehistoric far-western North America. Pp. 470-494 in Vulture Biology and Management, S.R. Wilbur and J.A. Jackson, eds. Berkeley, Calif.: University of California Press.

Skelton, P. 1993. A Complete Guide to the Freshwater Fishes of Southern Africa. Halfway House, South Africa: Southern Book Publishers.

Solz, D.L., and R.J. Naiman. 1978. The Natural History of Native Fishes in the Death Valley System. Sci. Ser. Publ. 30. Natural History Museum of Los Angeles County, Los Angeles, Calif.

Stafford, T.W., Jr., P.E. Hare, L.A. Currie, A.J.T. Jull, and D.J. Donahue. 1990. Accelerator radiocarbon dating at the moelcular level. J. Archaeol. Sci. 18:35-72.

Steadman, D.W. 1989. Extinction of birds in eastern Polynesia: A review of the record, and comparisons with other Pacific island groups. J. Archaeol. Sci. 16:177-205.

Steadman, D.W. 1991. Extinction of species;. Past, present, and future. In Global Climate Change and Life on Earth. New York: Routledge, Chapman, and Hall.

Steadman, D.W. 1992. Extinct and extirpated birds from Rota, Mariana Islands. Micronesica 25:71-84.

Steadman, D.W. 1993. Biogeography of Tongan birds before and after human impact. Proc. Natl. Acad. Sci. USA 90:818-822.

Steadman, D.W. 1995a. Ecosystem integrity: The impact of traditional peoples. Pp. 633-647 in Encyclopedia of Environmental Biology, volume 1. San Diego, Calif.: Academic Press.

Steadman, D.W. 1995b. Prehistoric extinctions of South Pacific birds: Biodiversity meets zooarchaeology. Science 267:1123-1131.

Steadman, D.W. and P.S. Martin. 1984. Extinction of birds in the late Pleistocene of North America. Pp. 466-477 in Quaternary Extinctions, P.S. Martin and R.G. Klein, eds. Tucson, Ariz.: University of Arizona Press.

Steadman, D.W. and N.G. Miller. 1987. California condor associated with spruce-jack pine woodland in the late Pleistocene of New York. Quat. Res. 28:415-426.

Steadman, D.W., T.W. Stafford, Jr., D.J. Donahue, and A.J.T. Jull. 1991. Chronology of Holocene vertebrate extinction in the Galápagos Islands. Quat. Res. 35:126-133.

Sublette, J.E., M.D. Hatch, and M. Sublette. 1990. The Fishes of New Mexico. Albuquerque, N.M.: University of New Mexico Press.

Swetnam, T.W. 1993. Fire history and climate change in giant sequoia groves. Science 262:885-889.

Turner, A. 1993. Species and speciation. Evolution and the fossil record. Quat. Int. 19:5-8.

Webb, T. III, P.J. Bartlein, S. Harrison, and K.H. Anderson. 1993. Vegetation, lake levels, and climate in eastern United States since 18,000 yr B.P. Chap. 17 in Global Climates Since the Last Glacial Maximum, H.E. Wright, T. Webb III, and J.E. Kutzbach, eds. Minneapol, Minn.: University of Minnesota Press.

Wilson, E.O. 1992. The Diversity of Life. Cambridge, Mass.: Harvard University Press.

Witte, F.T.J. Goldschmidt, J. Wanink, M. van Oijen, K. Goudswand, E. Witte-Maas, and N. Bouton. 1992. The destruction of an endemic species flock: Quantitative data on the decline of the haplochromine cichlids of Lake Victoria. Environ. Biol. Fishes 34:1-28.

3

Species Definitions and the Endangered Species Act

Many different cultures have an extensive literature on the history of the use of the term *species*, or kinds of organisms. The Endangered Species Act defines *species* in Section 3(15) to include "any subspecies of fish or wildlife or plants, and any distinct population segment of any species of vertebrate fish or wildlife which interbreeds when mature."[1] The act uses *species* in a legal sense to refer to any of those entities. However, *species* has vernacular, legal, and biological meanings.

The committee has been charged with examination of the use of the term *species* as it has been interpreted for implementation of the ESA from 1973 to the present. Since the ESA became law in 1973, there have been changes in what many biologists consider the term *species* to represent, and technological and philosophical innovations to find species. Nonetheless, the committee concludes that the ESA's inclusion of species, subspecies, and distinct population segments is correct and appropriate.

---

[1]However, the term *endangered species* excludes "members of the Class Insecta determined by the Secretary to constitute a pest whose protection under the provisions of this Act would present an overwhelming and overriding risk to man" (Section 3(6)).

46

HISTORICAL USE OF THE TERM *SPECIES* IN IMPLEMENTATION OF THE ENDANGERED SPECIES ACT

Introduction

Many societies have names for kinds of organisms, usually organisms that are large and conspicuous, or of life-sustaining, life-threatening, or economic importance. The term *species* can be applied to many of those kinds and be an accurate scientific term as well as an accurate vernacular term, because the characteristics used to differentiate species can be the same in both cases. Largely for this reason, defining what a species is has not been a major source of controversy in implementation of the Endangered Species Act. Greater difficulties have arisen in deciding about populations or groups of organisms that are genetically, morphologically, or behaviorally distinct, but not distinct enough to have been recognized at the rank of species according to traditional criteria—i.e., subspecies, varieties, and distinct population segments. Therefore, an important part of this chapter deals with taxa identified at ranks below the rank of species.

Zoological Interpretations of *Species*

Animals have always been a particular concern of the ESA—the imperilment of large, conspicuous animals in large part drove the popular demand for passage of the act. And because they are relatively well known from scientific as well as popular standpoints, birds and mammals have captured most of the concern, research, management, and funding associated with the ESA. However, these two groups of vertebrates constitute much less than 1% of animal species. The initial focus of ESA implementation on birds and mammals was logical and reasonable, given the general state of knowledge about biodiversity 20 years ago, but today implementation of the ESA continues to focus on a relatively small portion of the imperiled biota of the United States. Remedying this situation is an enormous challenge, given the poor state of our knowledge about many groups of invertebrates. However, of the 160 taxa of animals listed or proposed for listing from 1985 through 1991, 94 (59%) were vertebrates (Wilcove et al., 1993), and of those, only 38 (40%) were birds or mammals, showing that the bias perhaps is beginning to recede.

Before 1973, federal laws intended to preserve species applied only to native animal species. Passage of the ESA in that year extended coverage to all plants and invertebrate animals, bringing U.S. policy into line with the Convention on International Trade in Endangered Species of Wild Fauna and Flora (CITES), which also was ratified in 1973. Although plants and invertebrate animals were originally given equal status with vertebrates, subsequent emendation in 1978 restricted the use of distinct population segments to vertebrate animals only. Our survey of vertebrate listings indicates that, with few exceptions, species and subspecies have been almost exclusively listed; as discussed by Wilcove et al. (1993), only 2% of listings have been for populations.

Our examination of listing documents reveals that morphological features, such as color pattern, shape, scale patterns, and numbers of body elements, are overwhelmingly used for differentiation of taxa to be considered for protection. In part, this reflects the level of knowledge we have about the biota—practically no other information is available for most animals. In some more conspicuous or commercially valuable species, other factors are useful, such as breeding times or genetic analyses of winter-run populations of the chinook salmon, *Oncorhynchus tshawytscha* (*Fed. Reg.* 52:6041-6048).

Although protection of invertebrates is not extended to population levels in the ESA, subspecies are protected. Subspecies of invertebrates have merited some attention, especially in well-studied groups like mollusks (e.g., several subspecies of the pearly mussel *Epioblasma* (*Fed. Reg.* 41:24064-24067)), beetles (e.g., the valley elderberry longhorn, *Desmocerus californicus dimorphus* (*Fed. Reg.* 43:35636-35643) and the northeastern beach tiger beetle (*Fed. Reg.* 55:32088-32094)), and butterflies (e.g., the Oregon silverspot, *Speyeria zerene hippolyta* (*Fed. Reg.* 45:44935-44938)). The rare use of subspecies in many invertebrate groups probably reflects our general ignorance of those organisms more than any other factor. Overwhelmed by the vast numbers of taxa they must deal with, specialists have not had the time or resources to pursue the finer levels of variation in most groups. In each of the cases just cited, only morphological features were considered in differentiating subspecific taxa.

Botanical Interpretations of *Species*

When Congress passed the Endangered Species Act in 1973, it explicitly included plants, as well as fish and wildlife. After the ESA was passed, Congress requested that the Smithsonian Institution prepare a list of threatened, endangered, and extinct plants of the United States. In collaboration with botanists across the country, the Smithsonian Institution completed its list and presented it to the Congress in December 1974 (U.S. Congress, 1975). The list contained 2,832 endangered and threatened taxa and 355 presumed extinct taxa. The list contained only ferns, gymnosperms, and angiosperms. Of the endangered and threatened taxa, 1,999 were from the mainland of the United States and 833 were from Hawaii. The report recommended that the secretary of the interior review the Smithsonian's roster and publish lists of proposed endangered and threatened plants in the *Federal Register*.

After the initial involvement of the Smithsonian Institution, the Department of the Interior and the Department of Commerce took over responsibility for plants and animals. Responsibility for land and freshwater plants was accepted by the Fish and Wildlife Service (FWS) within the Department of the Interior, and the National Marine Fisheries Service (NMFS) in the Department of Commerce took responsibility for any marine plants proposed to be listed as threatened or endangered. To date, no marine plants have been proposed for listing. The director of FWS announced in 1975 that he intended to review the Smithsonian's list and determine eligibility of taxa for listing (*Fed. Reg.* 40:27823-27924). In 1976, FWS proposed that 1,726 of the taxa on the Smithsonian's register be listed (*Fed. Reg.* 41:24523-24572), but no action was taken on the recommendation. In 1978, Section 4(f)(5) of the Endangered Species Act was amended to state that species on which action had not been taken during 2 years after the initial proposal must be withdrawn from consideration. A grace period of 1 year was provided. By late 1979, action had still not been taken on the 1,726 plant taxa, and the list was withdrawn in December 1979 (*Fed. Reg.* 44:238). Nevertheless, by that time 47 plant taxa had been listed, establishing a precedent for placing plants on the endangered species list. As of January 31, 1995, 516 plant taxa (484 flowering plants, 4 conifers and cycads, 26 ferns and fern allies, and 2 lichens) had been listed, all but three occurring in the United States (George Drewry, FWS, pers. commun., Feb. 28, 1995), and many are under consideration.

The original Smithsonian report provided several additional recommendations and discussions. Under Recommendation 8, it was noted that the act provided differently for wildlife and fish compared with plants, in that the term *species* as applied to plants included subspecies but not varieties. The report also noted that the secretary of the interior did not have the authority to acquire land for conserving rare and endangered plant species unless they were listed in the appendices to CITES. Finally, in contrast to its provisions concerning animals, the act did not prevent the "taking" of endangered and threatened plant species in the United States.

Notwithstanding the Smithsonian report's comments about the use of the term *species* with regard to plants, the general practice in implementing the ESA has been to use the word *subspecies* in an unconventional sense to apply to either a subspecies or a variety. Despite their referring to different taxonomic categories (varieties are one rank below subspecies), the two words have effectively been used interchangeably for purposes of the act. Of the 404 plant taxa listed by 1993 [50 CFR 17.11 & 17.12, 1993], 342 are full species, 24 are subspecies, and 38 are varieties. Which infraspecific rank has been used tends to reflect the school of the taxonomists working on the listed group of plants (see Stuessy, 1990, for a historical review of the use of terms). It should be noted, however, that in the act, the word *species* is used for any category eligible for listing.

An analysis of the use of the term *plants* in the ESA and of the way in which categories have been chosen for protection under the act brings up several points. First, the term *plants* was specifically stated to include any member of the plant kingdom (ESA §3(13)) and was obviously meant to refer to any macroorganism that was not an animal—vascular plants, bryophytes, lichens, fungi, and so on. It seems most likely, however, that Congress did not intend to include most prokaryotes and many single-celled eukaryotes that might come under the rubric of plants (e.g., bacteria and yeasts), although botany departments classically taught mycology, microbiology, and phycology as well as seed-plant biology. Only two plants other than ferns, gymnosperms, and angiosperms have been listed (two lichens). However, there is growing recognition that the living world is not divisible into plant and animal kingdoms, and various schemes with several kingdoms have been proposed (e.g., Margulis and Schwartz, 1987). As currently written, the act could be construed as excluding a large portion of the living world, but lacking scientific knowledge about how species concepts apply in practice to many of these organisms, there is no basis for recommending the extension of the ESA's coverage to prokaryotes and many single-celled eukaryotes at this time.

Plants present problems in use of the concepts of species, subspecies, and varieties. When the Endangered Species Act was passed, the biological species concept of Mayr (Mayr, 1942; 1963), Stebbins (1950), and Grant (1963) was accepted by many systematists in theory, even though it often was difficult to apply. It was particularly difficult to apply to plants, which often exhibit polyploidy, hybridize with some frequency in nature, and frequently reproduce asexually or apomictically (without exchange of gametes among different individuals).

Discussions in several final listing documents for endangered species illustrate the ways species have been delimited by plant systematics. In the discussion proposing listing the fern *Thelypteris pilosa* var. *alabamensis* (*Fed. Reg.* 57:30165), eligibility depended in part on whether it was distinct from *T. pilosa* in Mexico. Distinctiveness was determined by asking experts to provide evidence of uniqueness. The consensus of experts was that several characteristics of the leaves were distinct, and therefore, the Alabama plants were recognized as an independent taxonomic variety.

A similar case involved *Sarracenia rubra* subsp. *alabamensis*, a pitcher plant (*Fed. Reg.* 54:10150). This taxon is part of a complex of populations that reproduce both sexually and vegetatively, hybridize, and exhibit a range of variation that is determined by local environment. Taxonomic recognition of the *alabamensis* subspecies was based on the unique possession of several leaf (pitcher) features in the Alabama individuals.

Different criteria were used to designate the Malheur wire lettuce, *Stephanomeria malheurensis*, as a species and its eligibility for listing. This taxon derives from *S. exigua*, but unlike its parent, it reproduces primarily by inbreeding and differs genetically from its parent by one fixed allele. Furthermore, chromosomal structural differences between *S. exigua* and *S. malheurensis* prevent successful meiotic pairing in the offspring. In the first two instances mentioned above, primarily morphological criteria were used to delimit a taxon. In the last case, genetic data and evidence of reproductive isolation were cited as evidence for specific rank.

The use of different kinds of evidence in different cases reflects the widely varying degrees of knowledge available for diffierent kinds of plants. Such variation should be expected for the foreseeable future, for plants as well as for vertebrate and invertebrate animals.

HISTORY OF SPECIES CONCEPTS BEFORE AND AFTER THE ENDANGERED SPECIES ACT

Species Concepts

Although Darwin's *Origin of Species* was ostensibly about speciation, the book really is about natural selection, or the changes that take place within species, not about how species arise from other species. Many biologists have extrapolated from Darwin's assumptions about selection to a theory of speciation. But biologists with different perspectives and problems in mind have different ideas about what a species is and what role it should play in particular areas of science. Some systematic biologists have declared that there is no single unit that can be called species, and, for example, that the concept of species used in classifying mosses might be quite different from that used for classifying species of birds with respect to population and genetic structure (Mishler and Donoghue, 1982; Mishler and Brandon, 1987; O'Hara, 1993). Such authors question why we should expect that some "unit of nature" equally meets the expectations and needs of ecologists, behaviorists, biochemists, and other kinds of biologists, and they suggest that noncongruence among species delineated using different methods often exists. This is equivalent to saying that various characteristics of organisms and interbreeding populations might change at different rates, so that differences in morphology, biochemistry, chromosome structure, breeding characteristics, and population structure, for instance, need not all arise together during differentiation.

Why should the term *species* be so problematic? Why, after centuries of investigations, are systematic biologists unable simply and easily to tell

us which groups of organisms are species and which are not? The reason is
that in the vast majority of cases, speciation is very gradual from a human
perspective, taking from hundreds to thousands of years or more. We believe that speciation is occurring in many kinds of organisms, but in the
extremely short time frames available to us, we cannot see the entire process unfold; we can see only what appear to be various lineages at different
stages in the process.

The most basic elements of speciation are isolation and differentiation.
For a simple case, imagine populations of a widespread species becoming
isolated due to a rising sea level that turns a small mountain range into an
island chain. Under current views of evolution, populations on each island
mountain-top would be expected to become different over time, and eventually to lose their ability to interbreed with each other. Whether *species* is
defined based on differentiation or loss of interbreeding, the isolates would
eventually become different species. The scenario could be expanded to
include several ancestral species, each widely distributed over the mountain
range, and thus with isolates on each island. On any particular island, the
degree of differentiation of the several kinds of organisms could vary, so
not all the descendent populations would become species at the same time.
Any given island might have entities that systematists would recognize as
full species, subspecies, distinct population segments, or even the ancestral
species, unaltered. To complicate matters further, a drop in sea level might
allow all these isolates to meet—those that had differentiated sufficiently
would not interbreed and would thus remain isolated as species, but others
might interbreed, spreading their special characteristics into the other populations so that the unique identities of each isolate would be lost.

Calling a group of organisms or populations a species or some other
rank is a hypothesis about the past—about differentiation, population biology, and history. Speciation is a dynamic process with many factors involved, and science cannot delineate what nature itself does not. That
complexity is a factor when competent scientists disagree about the status
of a particular group of organisms. We cannot know what is going to
happen in the future, but theories do give us models in which we can
examine many assumptions about speciation, and these are the best available tools for evaluating various parts of the genealogical nexus with respect to their conservation value.

Another vexing issue is whether species should or must be defined in
terms of the evolutionary processes thought to result in speciation; or whether
they should be defined based on their features alone. Species concepts
incorporating details of the evolutionary process are the "biological," "recognition," and some versions of the "evolutionary." The species concepts
that rely primarily on traits are the "cladistic" and "phylogenetic" (see Box
3-1). In part the debate is about the extent to which species should be

---

BOX 3-1
Species concepts in the scientific literature

Biological (BSC): Groups of interbreeding natural populations that are reproductively isolated from other such groups (Mayr, 1969; Mayr and Ashlock, 1991).

Cladistic: The set of organisms between two speciation events (Ridley, 1989).

Cohesion: The most inclusive group of organisms having the potential for genetic and/or demographic exchangeability. This is much like the BSC, but includes nonsexually reproducing individuals by emphasizing processes that keep populations from changing (e.g., ontogenetic constraint) (Templeton, 1989).

Evolutionary (ESC): A single lineage of ancestor-descendant populations that maintains its identity from other such lineages and has its own evolutionary tendencies and historical fate (Wiley, 1981).

Phylogenetic: The smallest aggregation of populations (sexual) or lineages (asexual) diagnosable by a unique combination of character states in comparable individuals (Nixon and Wheeler, 1990).

Recognition: The most inclusive population of individual biparental organisms that share a common fertilization system (Paterson, 1985).

---

construed as units of evolution or whether evolution should be envisioned as acting at other levels (Kluge, 1990). There is a vast literature on all these topics, including papers by Cracraft (1987), Davis and Nixon (1992), and Ereshefsky (1989) and by Otte and Endler (1989), Paterson (1993), and de Queiroz and Donoghue (1988).

Developments in population genetics since the 1920s have led systematists to focus on populations of organisms. Most species concepts began to incorporate the view that the genetic interactions within species are the major factor in maintaining species identity, and loss of those interactions is the major factor in speciation. The biological species concept was adopted generally among zoologists studying vertebrates, and to a lesser extent, among botanists and zoologists studying invertebrates. Many of the debates about the biological species concept have concerned its application to organisms without sexual reproduction.

Indeed, the reproductive differences between plants and animals have fueled the different uses of species concepts among disciplines. Incorporation of genetics into plant systematics in the 1940s and 1950s led to the use of degrees of crossability as indicators of relationships (Camp and Gilly, 1943; Grant, 1957). This approach paralleled the rise and domination of the biological species concept in zoology, and in time contributed to a merging of the two. The adoption of the biological species concept ushered in an era of "biosystematics" in botany that relied heavily on the degree of interfertility as a guide to degrees of relatedness (Gilmour and Heslop-Harrison,

1954; Grant, 1959; Solbrig, 1970; Stebbins, 1970; Raven, 1974). Inability to interbreed was considered evidence of distinctiveness at the species level. However, despite this research in biosystematics, most plant systematists proceeded without a methodological or philosophical conviction that the biological species concept was applicable to most plant species. It was obvious that some entities ranked as genera on morphological and ecological grounds could cross in nature, whereas taxa judged to be nearly identical on morphological grounds might be completely reproductively isolated from one another. In addition, plants are often completely functionally or truly asexual in their reproduction and therefore are excluded even by Mayr (1942) from a definition that relies on interbreeding. The ability of an organism to interbreed is not always a safe criterion for species distinctiveness. Some organisms that are distantly related might interbreed because their reproductive systems have not differentiated.

In 1974, Ghiselin's "radical solution to the species problem" restructured the debate to a consideration of the ontological status of those units of nature being called species. Ghiselin's timing was coincident with the establishment of phylogenetics in systematics after 1966, and the general rethinking of many basic assumptions that took place in those years included reexamination of species concepts as well. In phylogenetic systematics, many biologists found an approach that advocated the recognition of species as distinct lineages of differentiated organisms. Species are defined as units that are diagnosable (are distinctive) and that have a unique evolutionary role or trajectory (Wiley, 1981; Donoghue 1985; Mishler, 1985). In addition, spatiotemporal restriction for species is expected, i.e., individuals of a species will share the same broad geographic range and temporal range. This prevents similar-looking individuals that have evolved from different ancestors in different parts of the world (or in different geological times), from being considered conspecific. In adopting the phylogenetic species concept, many scientists have moved to a species concept very much like the inclusive one used in the ESA. Indeed, despite differences in theory and philosophy underlying various species concepts, systematists using different concepts usually come to substantial agreement with respect to the recognition of species, especially in vertebrates and in many higher plants.

Taxonomic Units Below the Rank of Species

Within the phylogenetics community, the usefulness or validity of recognition of units below the species level under a phylogenetic species concept (de Queiroz and Donoghue, 1988; Avise and Ball, 1990; Nixon and Wheeler, 1990; Davis and Nixon, 1992; Waples, in press) has been the subject of debate. One argument is that any distinctive population or group of populations should be considered a species. Because subspecies and varieties are recognized by one or a suite of characters, they are certainly distinctive. However, the evolutionary independence of such units (whether there is a significant amount of interbreeding among them) can be disputed, as can whether their recognition as separate species obscures the degree of relationship among them.

Despite of the lack of consensus about species concepts, the scientific and popular cultures attach special significance to the rank of species, leading to debates over the importance of groups of organisms ranked below the species level, such as subspecies, varieties, races, or population segments. Former Secretary of the Interior M. Lujan once asked whether we have to save every subspecies (Lancaster, 1990). The implicit assumption was that species have greater value than lower ranks. In practice, recent application of the law has been primarily to entities considered species by systematists. A review of listings for 1985-1991 found that only 18% of listed vertebrate taxa were subspecies, and only 2% were distinct population segments (Wilcove et al., 1993).

Missing from most discussions about rank of taxa to be preserved is the crucially important recognition that different concepts of species, subspecies, and other ranks are often applied between and even within disciplines. Many of the subspecific and population-level taxa in the 18% just mentioned were birds. In part, this is due to the ornithological tradition of recognizing certain kinds of variation at "subspecific" rather than "specific" level. A fish biologist looking at a similar kind of variation might well have used the species rank in describing what an ornithologist would consider a subspecies. Even within a discipline, recognition of degrees of variation can vary over time, among systematists subscribing to different schools, and between practitioners from different countries. In addition, in phylogenetics or cladistics, rank is de-emphasized. Some phylogenetic classifications eschew rank altogether, except for ranks required by codes of nomenclature (e.g., every species must be placed in a genus). Thus, since ranking of a taxon may vary for any number of reasons, we find the ESA's inclusion of all three categories for preservation—species, subspecies, and distinct population segments (at least of vertebrates)—correct and appropriate. Moreover, there is no scientific reason that the ESA's inclusion of all three ranks should not apply to all groups of organisms. But Secretary Lujan's question about preservation of all subspecies has relevance for policy makers. Unless we agree to preserve all endangered or threatened organisms of all taxonomic ranks, we must find ways to identify those groups of organisms we consider to be significant.

A CONCEPT OF SPECIES FOR THE PURPOSES OF THE ENDANGERED SPECIES ACT

Introduction

After centuries of debate, no one doubts that natural groups of organisms exist. Scientists now are concerned with methods of classification and circumscription of taxonomic boundaries. For the purposes of the ESA, whether an entity is a species in the vernacular sense is less important than whether the entity is a group of individuals that can be held to be distinct from other such groups, with all that such distinctiveness implies regarding population structure and evolutionary potential. Fortunately, the groups of organisms called species and subspecies by taxonomists usually are also distinct in their population structure and evolutionary potential. The most difficult questions generally arise at taxonomic levels below the subspecies level. Because evolutionary units at such levels are not discrete but exist along a continuum, it is a policy judgment as well as a science judgment to determine the significance of an evolutionary unit. In other words, science alone does not lead to a conclusion that any objectively definable degree of distinctiveness is more significant than another. But science can provide the tools for identifying and measuring biological components of distinctiveness; it also can provide objective criteria for ranking the distinctiveness of evolutionary units.

Evolutionary Units and Their Identification

The following provides some guidelines for measuring the distinctiveness of natural entities to serve the purposes of the ESA. The committee focused on the biological meaning and evolutionary importance of distinctiveness. To be clear on this matter, the committee considered what distinctiveness means separately from considering how to assess whether a particular population is in need of protection. The committee believes that separating the two issues—as the ESA does—would help policy makers and managers as well. What follows will bring increased scientific objectivity to the current and appropriate case-by-case examination of whether a given population or taxon below the rank of subspecies is eligible for protection by the ESA. It should also help to meet Congress's challenging expectation that distinct population segments be listed "sparingly and only when biological evidence indicates that such action is warranted" (S. Rep. 151, 96th Congress. 1st Session, 1979).

Species and Subspecies

The ESA is clear that species and subspecies of fish, wildlife, or plants—defined in the act to include all members of the plant and animal kingdoms—are eligible for protection. Given the intent of the ESA to protect biological diversity, the coverage in the ESA has a sound scientific basis. In practice, it is often necessary to rely on the judgments of competent systematists to identify taxa, especially subspecies.

Distinct Population Segments

To help in identifying distinct population segments we developed the concept of an evolutionary unit (EU).[2] An EU is a group of organisms that represents a segment of biological diversity that shares evolutionary lineage and contains the potential for a unique evolutionary future. Its uniqueness can be sought in, but is not limited to, attributes of morphology, behavior, physiology, and biochemistry. A basic characteristic of a EU is that it is distinct from other EUs. In most cases, an EU will also occupy a particular geographical area. Clearly, most currently recognized species and subspecies are also EUs. Thus, the following discussion provides a practical approach to identifying distinct population segments.

Application of the evolutionary unit concept should result in no substantive change in the application of conservation laws. The committee believes its use will move decisions of eligibility for protection away from arguments about taxonomic ranks and into a realm where more substantive views about the degree to which populations are evolutionarily significant and new techniques can be applied.

The term *distinct population segment* has been applied two ways in implementing the ESA. One way is that the distinct population segment corresponds to an EU—an evolutionarily distinct population segment that is geographically or otherwise isolated from other population segments. In some cases, the population segment being referred to is a metapopulation, i.e., a larger population made up of smaller, local breeding populations that have some genetic and ecological interactions among them. A good ex-

---

[2]The EU is similar but not identical to the "evolutionarily significant unit" (ESU) proposed by NMFS for managing anadromous Pacific salmonids (NOAA, 1991; Waples, 1991; Waples, in press). The EU does not stress reproductive isolation as a criterion, because reproductive isolation is often difficult to assess directly, and it stresses evolutionary future more explicitly than the ESU does. By omitting the word "significant," it recognizes that significance is a continuum, a point clearly expressed also by Waples (in press). However, it seems likely that the application of either the EU or the ESU concept would lead to similar results most of the time, especially for vertebrates.

ample of a metapopulation is salmon in a large river system. In this case (e.g., winter-run chinook salmon in the Sacramento River of California, listed as endangered), the local breeding populations are the fish that spawn in each small tributary stream. A recent NRC report (NRC, 1995) described the metapopulation concept as it applies to salmon in great detail, and Waples (in press) discussed the applicability of the ESU to metapopulations. Although identifying an EU might be more difficult if the population in question is a metapopulation, it is no different in concept from identifying a single population as an EU.

Political boundaries also have been used to designate population segments. Although those boundaries can be appropriate to delineate population segments for management, they often do not delineate EUs, especially when the political boundaries do not coincide with major natural geographic boundaries. Although there can be good policy reasons for such delineations, there are not sound scientific reasons to delineate species only in accordance with political boundaries.

Distinctiveness

The fundamental characteristic of an EU is that it is distinct from other EUs, and distinction implies an independent evolutionary future. Estimates of distinctiveness (circumscription of EUs) are based on genetic, molecular, behavioral, morphological, or ecological information. But a single kind of information often will fail to provide compelling evidence of distinctiveness. Determination of distinctiveness and the associated inference of an independent evolutionary future usually requires the careful integration of several lines of evidence. In this section, we discuss some of the experimental and observational tools currently used to assess distinctiveness.

Genetic Isolation

Because possession of a unique potential for evolution is an important characteristic of an EU, genetic difference of a population from other populations is an important aspect of distinctiveness. It is also evidence for physical or reproductive isolation, because populations whose members interbreed cannot become completely differentiated genetically. Isolation over a prolonged period is usually associated with sufficient genetic divergence to ensure an independent evolutionary future. Measures of genetic divergence or genetic distance between populations can be used as surrogate measures of the degree of genetic isolation between them. A large degree of genetic divergence or genetic distance between two populations is prima facie evidence of distinctiveness. That is because even a small amount of interbreeding between two populations will reduce the genetic divergence between them unless there is strong differential selection between the two populations. If there is strong differential selection between two populations, they are likely to diverge genetically after a time.

The most direct test of isolation is testing for the ability of individuals from different populations to interbreed. Genetic analysis also includes tests to see if differences in morphological or behavioral characteristics between populations have a genetic basis, and molecular methods for assessing genetic variation directly. Box 3-2 describes several of the methods currently in use for assessing variation in the genetic material and in the proteins that genes produce. The wide application of these methods (particularly the isozyme method) to plant, animal, and microbial species has provided a general picture of the average genetic divergence associated with various levels of taxonomic distinction. However, average values cannot be applied to particular situations. Rather, distinction must be assessed on a case-by-case basis.

Many species are composed of a network of populations (a metapopulation system) distributed across a heterogeneous physical environment. Understanding this aspect of population structure is important for protecting some endangered populations, because a particular population of concern might depend for its long-term survival on occasional interbreeding with other populations in the metapopulation of which it is a part. Particular populations within a metapopulation can exchange genes through migration. But barriers to genetic exchange ultimately lead to genetic isolation, and when genetic exchange is at a sufficiently low level, populations can be effectively isolated. As a result, they and perhaps the whole metapopulation might be more vulnerable to extinction.

Geographic and Temporal Isolation

Many methods help to detect geographic isolation between populations. One technique uses tags or markers to follow individuals to see if they move into areas occupied by other populations. Some tags that have been used are natural morphological variations, a variety of artificial marks, and natural molecular or isozyme markers. An interesting example of isolation is provided by pink salmon (*Oncorhynchus gorbuscha*) of the Pacific coasts of North America, Russia, and Japan. Pink salmon have rigid 2-year life cycles, so fish spawned 1 year apart (referred to as "odd-year" and "even-year" fish) almost never interbreed, even though they spawn in the same streams (Heard, 1991).

Behavioral and Reproductive Isolation

Populations of many organisms are isolated from each other because of

BOX 3-2

Enormous advances in genetic technologies have been made since the ESA was first passed in 1973. The most widely applied technology is the isozyme method. This method permits the detection of genetic differences in enzymatic proteins based on the physical separation of protein molecules in a supporting medium (usually some form of gel, such as starch or polyacrylamide). Because all protein molecules are electrically charged (under defined pH conditions), they can be physically separated by applying an electric field to a gel for a specified period. The distance migrated is a function of the resistance of the protein in the gel (largely a function of protein size) and the net charge of the protein (the larger the net charge differential, the greater the rate of migration). Roughly one-third of all mutational differences among individuals will cause a change in a particular amino acid in protein coding genes. A subset of amino acid substitutions have the property of altering the net charge of a protein. Consequently, a small fraction of all mutations in protein coding genes can be detected as changes in the mobility of the protein in an electric field. The final step in the isozyme method is the application of a histochemical stain to the gel to make the sites of enzymatic activity visible. Histochemical stains contain the substrate for a specific enzyme (e.g. alcohol for alcohol dehydrogenase) and a coupling agent that reacts with the product of the enzymatic reaction to cause the precipitation of a dye at the site of enzyme activity. Genetic differences among individuals are read as differences in banding locations on the gel.

At present, nearly 100 different enzymatic proteins can be resolved in humans. About 25 to 30 are routinely resolved in most plant and animal species. As a consequence, average levels of genetic divergence can be estimated from reasonably large samples of genes that determine protein products. In addition, the isozyme method is relatively easy to implement, and it is not difficult for an experienced laboratory to screen several hundred individuals per day. Because of ease of application and because the isozyme method was invented more than 30 years ago, it has been widely applied in population genetics and in systematics. A large data base has been accumulated that spans hundreds of plant and animal species.

The statistics calculated from isozyme data are numerous, but most involve estimating the gene frequencies for a particular genetic locus and then calculating a measure of genetic diversity averaged over loci (Weir, 1990). Gene frequencies are estimated by counting the number of copies of a particular allele in the sample and then, for diploid organisms, dividing by twice the number of individuals in the sample (recall that every diploid individual received one copy of a particular gene from each parent, and hence there are twice as many genes as individuals in the sample). The most commonly employed diversity measures are based on calculating the average probability of drawing two different copies of a gene at a locus. This measure is then averaged over loci. A large body of population genetic theory relates diversity measures to processes of drift, mutation, and migration (see, e.g., Nei, 1987; Weir, 1990).

"Molecular methods" usually refer to the set of methods that involve direct assay of either DNA or RNA. A large number of techniques fall under this

BOX 3-2 Continued

rubric. Moreover, there continues to be a rapid rate of innovation in molecular technology. Accordingly, we restrict our comments to technologies involving comparisons of DNA sequences or patterns of restriction fragments. Both these technologies now rely heavily on the use of PCR (polymerase chain reaction) methods that permit the rapid and accurate generation of large amounts of DNA for the gene or molecular region of interest. The PCR method was invented in the mid-1980s, and it is now beginning to be widely applied in population genetics and systematics. The PCR method depends on some prior knowledge of the DNA sequence of the gene of interest for the design of primers for the DNA polymerase enzyme (except for the RAPD application, which is discussed below). The primers together with a heat stable DNA polymerase and a thermal cycling apparatus allow the geometric amplification of the DNA region of interest (a region of up to roughly 4,000 to 5,000 base pairs in size might be amplified).

Once a DNA fragment has been amplified by PCR, the fragment size can be compared among samples using gel electrophoresis, or the fragment can be further analyzed by more sophisticated methods. Changes in fragment size imply mutational changes resulting from the insertion or deletion of DNA segments. Among more sophisticated methods are (1) the direct sequencing of PCR products to identify particular mutations at the ultimate level of detail, the amino acid bases, or (2) further dividing the PCR-amplified fragment using a class of enzymes known as restriction endonucleases (restriction enzymes) to detect mutational changes in "restriction sites." DNA sequencing has been employed in several specific cases at the population level, but current technology still limits this application to relatively small samples. Restriction-site analysis is based on the fact that restriction enzymes cut double-stranded DNA molecules at precisely defined DNA sequences (e.g., GAATTC is the restriction site for the enzyme EcoR1), and a change in fragment pattern must therefore be due to a mutational change in a recognition site.

A recent innovation in the PCR method, know as the RAPD method, is based on short random primers (usually 10 nucleotides in length). PCR based on short random primers causes random regions in the genome bounded by sequences complementary to the primer to be amplified. This method is easy to implement and it is beginning to be widely applied in population biology; however, more limited genetic information is contained in RAPD analyses because banding patterns are usually not codominant (codominant means that in the case of heterozygosity, both gene products will appear on the gel as two distinguishable bands; in the case of dominance, only one band will appear, and thus less information can be obtained).

Current applications of PCR-based methods in endangered species management and research involve (1) the use of PCR in determining genealogical relationships to better manage breeding programs; (2) the use of PCR in law enforcement, where the taking of particular specimens can be established through forensic use of DNA; and (3) limited population surveys to establish genetic relationships among the elements of a metapopulation. Like the isozyme method, these other molecular methods help establish whether a population or system of populations is distinctive in the sense of constituting an EU.

differences in behavior or reproductive mechanisms. Populations of many fish species have very specific times and places where they spawn, so their spawning behavior prevents the populations from interbreeding. Differences in courtship behaviors—including songs and other vocalizations—often lead to reproductive isolation (e.g., Lack, 1971).

The degree to which population segments in the wild are independent units or part of a larger genetic entity is often unclear because of difficulties in ascertaining historical and current levels of gene flow. A key question is the extent to which the population segments have adapted to local environments in ways that would substantially reduce the probability of successful establishment in other portions of the species range. Tests for distinctiveness are most difficult to apply in cases where systems of populations are in the process of diverging to independent status. Because the process of evolutionary change is dynamic, it is inevitable that ambiguous cases will arise, especially as we further understand the complex population structures of some widespread species. As for any difficult and technical activity, the judgment of professional experts is essential. Despite such difficulties, the number of ambiguous cases is a small fraction of the total number of cases so far dealt with in the practical implementation of the ESA, and probably will remain so.

Examples of Circumscription of Evolutionary Units

Of the terrestrial vertebrate taxa that already have been listed under the ESA, the majority qualify as EUs. Few would doubt the taxonomic distinctiveness of such well-known animals as the American alligator, bog turtle, California condor, whooping crane, or black-footed ferret. Traditionally, rather subtle differences between closely related terrestrial vertebrate taxa, such as the various forms of grizzly (brown) bear and gray wolf, have been recognized at the subspecies or distinct population segment level. When such distinctions can be demonstrated as valid using any of the recognition criteria we have outlined, the taxon would qualify as an EU.

An example of an EU is the Allegheny woodrat, *Neotoma magister*, which occurs (or did until very recently) from southeastern New York and westernmost Connecticut through the foothills and higher elevations of the Appalachians to Kentucky, Tennessee, northernmost Alabama, and eastern North Carolina. In spite of morphological information on its distinctiveness (Goldman, 1910), the Allegheny woodrat had been considered conspecific with the more widespread Eastern woodrat (*N. floridana*) since the 1940s. However, recent studies have provided substantial additional evidence that *N. magister* is distinct from *N. floridana* in its morphology (Hayes and Richmond, 1993) and have demonstrated distinctiveness of mitochondrial DNA (Hayes and Harrison, 1992), as well as behavior and ecology (Wiley,

1980; Hayes and Harrison, 1992). These new data are pertinent to the ESA because, during the past several decades, the populations of *N. magister* have declined precipitously along the entire northern margin of its range. The species no longer occurs in Connecticut, New York, and New Jersey, and is rare in Pennsylvania, Ohio, Indiana, and Illinois. Bones from archeological and paleontological sites in New York and Indiana show that *N. magister* occurred in pre-European times even farther north than indicated by its range in the 19th and 20th centuries (Richards, 1971; Steadman, 1993a, b). The cause of the decline of *N. magister* is uncertain, although an ascarid roundworm transmitted via raccoon feces is likely to be involved (A. Hicks, New York Department of Environmental Conservation, pers. commun., Sept. 1993); because of events related to human activities (e.g., expansion of suburbs, loss of large predators, decline of hunting and trapping of raccoons), raccoon populations have expanded greatly in the past several decades. At the present rate of decline, *N. magister* may be eliminated from its entire range within the next 100 years.

Other examples of EUs are the distinct populations of otherwise western or tropical species found in peninsular Florida. The Florida populations (such as the burrowing owl, *Athena cunicularia floridana*) generally are distinct morphologically and traditionally have been recognized as subspecies endemic to Florida. The endemic Florida population of the scrub jay (*Aphelocoma coerulescens coerulescens*), which is declining because of habitat loss, differs from other populations genetically (Peterson, 1992) and in its behavior and ecology (Woolfenden and Fitzpatrick, 1984) and has recently been recognized by the American Ornithologists' Union's Committee on Classification and Nomenclature as a full species, *Aphelocoma coerulescens* (Richard Banks, president, AOU, pers. commun., 1995). Waples (in press) described in detail the identification of EUs (ESUs in NMFS terminology) of anadromous Pacific salmon, including some difficulties encountered.

Some kinds of genetic mutations occur commonly and sporadically in populations of organisms but do not usually result in speciation. Melanism is one of these. In many areas of the country, populations of squirrels (including the grey squirrel, *Sciurus carolinensis* and the fox squirrel, *Sciurus niger*) are either partially or primarily composed of black individuals. These populations do not qualify as EUs, because they are not isolated from adjacent populations that lack the gene for melanism or have it only at low frequencies. Similarly, melanism is a common mutation in the mosquitofish (*Gambusia affinis* or *G. holbrooki*). In this case, patterns of distribution indicate that dark individuals are the result of mutations occurring spontaneously throughout the range of the species. Even in cases where the melanism is known to be adaptive, as in the pepper moth (*Biston betularia*) described by Kettlewell (1961), the dark forms continue to interbreed with the light forms, so the dark forms do not constitute an EU. Similarly, a growing body of evidence suggests that different genetic variants at the major histocompatibility complex (MHC) loci in humans confer differential adaptation to disease, but we would not argue that the set of individuals who possess a particular MHC variant constitutes an independent evolutionary lineage, because such individuals continue to interbreed with other humans.

Other examples of organisms that would not be considered EUs are the American alligator (*Alligator mississippiensis*) populations that are regulated in the southern United States. Alligator populations may be regulated locally, with hunting allowed or protection bestowed based on political boundaries. The American alligator is an EU, but the regulated and unregulated populations are not EUs. The American brown bear (*Ursus horribilis*) is an EU (or perhaps several), but in many cases its protection status involves political boundaries and is based more on management and aesthetic criteria than on whether the populations are EUs or something like them.

More problematic is the status of populations of the American bald eagle of the contiguous United States. Those populations declined seriously before passage of the ESA, although Alaskan and Canadian populations remained large. In this case, especially because there is no discontinuity between the Canadian and U.S. distributions of the eagle, the EU is based on political boundaries. We cannot argue that a biological difference arises when a Canadian eagle flies across the border into the United States. The Florida panther (*Felis concolor coryi*) is another problematic case. Little genetic variation is evident in the Big Cypress panthers. Panthers in the Everglades, which were introduced from stock descended from Latin American populations, have hybridized with those in the Big Cypress area (Roelke et al., 1993; Barone et al., 1994). The Big Cypress population is so small and so lacking in genetic variation that it is unlikely to survive without further introduction of panther genes from elsewhere (Roelke et al., 1993; Barone et al., 1994). Florida panthers are no longer genetically distinct from other populations of panthers in the southeastern U.S., so the question becomes, as an FWS official put it to the committee, "whether we are protecting the Florida panther or protecting the panther in Florida."

In the case of the American bald eagle, and in other similar cases, there might well be persuasive reasons to conserve taxa distinguished by political jurisdiction (including ethical and aesthetic reasons). Large carnivores often play a major role in ecosystems, so there can be additional ecological arguments for protecting them. Others have argued that refusal to list only geographic segments of a species could deny protection to genuinely endangered populations and local ecosystems and might allow a widely distributed EU to decline until the entire species becomes endangered. This kind of argument has been applied in criticism of the application of the ESU concept (see Waples (in press) for a discussion of these and other criticisms), and it is a difficult aspect of the problem. A biologically sound method of identifying distinct population segments does not recognize political boundaries, although it does recognize the validity of asking whether a particular population within a political boundary is distinct and imperiled.[3]

This kind of criticism of an EU concept confounds identification of distinctiveness with a decision as to whether a particular population is in need of protection. Both are needed, but considering them separately helps the analyses. It is surely true that many listed species are endangered because widely distributed EUs were permitted to decline until they became imperiled, but this is not a flaw of the EU concept. Such a management strategy is bad for conservation, but the ESA and its regulations are intended to protect threatened and endangered species, not to prevent them from becoming threatened and endangered. Preventing species from becoming threatened and endangered is essential for preserving biological diversity, and additional conservation and management plans beyond the provisions of the ESA are needed to achieve that goal. Some of those are discussed in Chapter 10.

The Fate of Hybrids under the Evolutionary Unit Concept

Many organisms, especially plants, are descended from hybrids. In cases where the populations have become independent evolutionary units whose persistence *no longer* depends on hybridization among other such units, an EU can be recognized (Funk, 1985) if the population meets the criteria discussed above. In most zoological cases, hybrids will not be considered as EUs because they are temporary products of a breakdown in species boundaries and require continued contact and interbreeding among these species for their existence (Parsons et al., 1993). For example, the North American flickers (*Colaptes auratus* broadly defined) are widespread in North America and consist of three major types, each of which has been recognized by various authorities as distinct at either the species or subspecies level: the eastern yellow-shafted flicker (*C. auratus auratus*); the western and montane red-shafted flicker (*C. auratus cafer* broadly defined); and the southwestern gilded flicker (*C. auratus chrysoides*). Hybrids of *C. a. auratus*

---

[3]A draft joint FWS and NMFS policy on recognition of distinct vertebrate population segments under the ESA (*Fed. Reg.* 59:65884, 12/21/1994) incorporates this criterion. The policy has three elements: discreteness, significance, and status. One of the criteria for discreteness is the delimitation of a population "by international boundaries within which differences in control of exploitation, management of habitat, conservation status, or regulatory mechanisms exist that are significant. . ." In other words, the criterion leads to a question such as whether a population of organisms in the United States is distinct and endangered.

and *C. a. cafer* can be found along a north-south zone east of the Rocky Mountains, while hybrids of *C. a. cafer* and *C. a. chrysoides* are found locally in the southwestern United States and northwestern Mexico. Virtually everywhere they occur, all pure forms of *C. auratus* are common birds. Furthermore, the hybrids are common within much of their limited range and almost always occur alongside pure forms. Even the least common and most localized hybrid flicker (*C. a. cafer* x *C. a. chrysoides*) would not be considered an EU, because it is genetically dependent on the parent species.

The hybridization between the blue-winged warbler, *Vermivora pinus* (BW), and golden-winged warbler, *V. chrysoptera* (GW), is a more complex situation from the standpoint of recognition as an EU. The two species hybridize readily, producing viable offspring (Gill, 1980) that are sometimes called Brewster's or Lawrence's warblers. The GW is declining over much of its range (Confer and Knapp, 1992). Smith and coworkers (1993) assessed the status of 82 species of migratory birds in the Northeast and rated the GW on a par with the cerulean warbler (*Dendroica cerulea*) as the most jeopardized species. The decline is due mostly to a loss of habitat (the GW prefers thickets with small, scattered trees, a habitat that often is ephemeral due to succession), with other factors being brood parasitism by cowbirds, and hybridization with the BW. Where the GW and BW come together (in many localities in the Northeast and upper Midwest), the long-term (decadal) results of hybridization often favor the BW, with the GW becoming scarce or absent (Gill, 1980, 1985; Confer and Knapp, 1981). The BW prefers forest edges (especially with aspens), a type of habitat that is common today because of the patchiness of the second-growth forests. The GW on its own qualifies as an EU, but today's reality is that some percentage of phenotypic GWs share genetic material with the BW. Given that it is not feasible to assess the genetic make-up of all birds across a broad range, the phenotypic GWs should be recognized as an evolutionary unit in spite of the fact that some percentage of these birds contain BW genetic material. Hybrids that are phenotypically intermediate could not be diagnosed as part of the GW evolutionary unit.

The EU can easily be applied to hybrids. The principle would be to protect individuals that might have some introgressed genetic material (especially, for example, mitochondrial DNA) but that remain phenotypically much like the endangered parental species. Modern genetic techniques routinely find examples of interbreeding that has left the parent species' lineages essentially unchanged phenotypically.

Many developments in theoretical population genetics and in molecular technology since 1973 will frequently aid in resolving hybridization questions. They should play a central role in deliberations concerning decisions to list. This country has substantial expertise (in academia, museums, federal and state institutions, environmental organizations, and industry) on the systematics, population genetics, and ecology of representatives of most groups of plants and many animals, as well as a strong academic base in the broader fields of molecular evolution and genetic data analysis. A mechanism needs to be implemented allowing the federal government to take advantage of this expertise rapidly in a nonpolitical way and with the allocation of resources toward research and implementation. Decisions relevant to the ESA can be made more objectively than they have been, and this objectivity should ultimately reduce costs of implementing and enforcing the act by facilitating the decision-making process and reducing court costs.

CONCLUSIONS AND RECOMMENDATIONS

• The ESA's inclusion of species and subspecies is soundly justified by current scientific knowledge and should be retained. Often, competent systematists will be required to delineate subspecies and sometimes, species as well.

• The ESA's inclusion of distinct population segments—i.e., taxa below the rank of subspecies—is also soundly based on scientific evidence. To help provide scientific objectivity in identifying these population segments, the concept of the evolutionary unit (EU) should be adopted. We define the EU as a segment of biological diversity that contains a potential for a unique evolutionary future. Criteria to establish identity of an EU primarily concern assessment of its diagnosibility, or distinctiveness, relative to other units. Determinants of distinctiveness include morphology, behavior, genetics, molecular make-up, physiology, and so on. An analysis might include such factors as reproductive isolation, genetic variation, ecological distinctiveness and importance, details of reproductive ecology and dispersal, geographic isolation, and historic and prehistoric range changes and their causes. To clarify the analyses, identifying an EU should be separate from deciding whether it is in need of protection.

• The ESA explicitly covers species and subspecies of all plants and animals. As currently written, however, it covers taxonomic ranks below the subspecies level (distinct population segments) only for vertebrate animals. There is no scientific reason to exclude any EUs of invertebrate animals and plants from coverage under the ESA.

• Although the way organisms are divided into kingdoms has changed since the ESA was enacted in 1973, current scientific knowledge about how species concepts apply in practice to many of these organisms does not lead us to recommend that coverage be extended to prokaryotes and most single-celled eukaryotes, such as yeasts.

REFERENCES

Avise, J.C., and R.M. Ball. 1990. Principles of genealogical concordance in species concepts and biological taxonomy. Oxford Surv. Evol. Biol. 7:45-67.

Barone, M.A., M.E. Roelke, J. Howard, J.L. Brown, A.E. Anderson, and D.E. Wildt. 1994. Reproductive characteristics of male Florida panthers: Comparative studies from Florida, Texas, Colorado, Latin America, and North American zoos. J. Mammal. 75:150-162.

Camp, W.H., and C.L. Gilly. 1943. The structure and origin of species with a discussion of intraspecific variability and related nomenclatural problems. Brittonia 4: 323-385.

Confer, J.L., and K. Knapp. 1981. Golden-winged warblers and blue-winged warblers: The relative success of a habitat specialist and a generalist. Auk 98:108-114.

Confer, J.L., and K. Knapp. 1992. Golden-winged warbler *Vermivora chrysoptera*. Pp. 369-383 in Migratory Nongame Birds of Management Concern in the Northeast, K.J. Schneider and D.M. Pence, eds. U.S. Department of the Interior, Fish and Wildlife Service, Newton Corner, Mass.

Cracraft, J. 1987. Species concepts and the ontology of evolution. Biol. Philos. 2:329-346.

Davis, J.I., and K.C. Nixon. 1992. Genetic variation, populations, and the delimitation of phylogenetic species. Syst. Biol. 41:421-435.

de Queiroz, K., and M.J. Donoghue. 1988. Phylogenetic systematics and the species problem. Cladistics 4:317-338.

Donoghue, M.J. 1985. A critique of the biological species concept and recommendations for a phylogenetic alternative. Bryologist 88:172-181.

Ereshefsky, M. 1989. Where's the species? Comments on the phylogenetic species concepts. Biol. Philos. 4:89-96.

Funk, V. 1985. Phylogenetic patterns and hybridization. Ann. Mo. Bot. Gard. 72:681-715.

Ghiselin, M. 1974. A radical solution to the species problem. Syst. Biol. 23:536-544.

Gill, F.B. 1980. Historical aspects of hybridization between golden-winged and blue-winged warblers. Auk 97:1-18.

Gill, F.B. 1985. Whither two warblers? Living Bird Q. 4:4-7.

Gilmour, J.S.L., and J. Heslop-Harrison. 1954. The deme terminology and the units of microevolutionary change. Genetica 27:147-161.

Goldman, E.A. 1910. Revision of the wood rats of the genus *Neotoma*. North Am. Fauna 31:1-124.

Grant, V. 1957. The plant species in theory and practice. Pp. 39-80 in The Species Problem, E. Mayr, ed. American Association for the Advancement of Science, Washington, D.C.

Grant, V. 1959. Natural History of the Phlox Family. Systematic Botany, Vol. 1. The Hague, Netherlands: Nijhoff.

Grant, V. 1963. The Origin of Adaptations. New York: Columbia University Press.

Hayes, J.P. and R.G. Harrison. 1992. Variation in mitochondrial DNA and the biogeographic history of woodrats (*Neotoma*) of the eastern United States. Syst. Zool. 41:331-344.

Hayes, J.P. and M.E. Richmond. 1993. Clinal variations and the morphology of woodrats (*Neotoma*) of the eastern United States. J. Mammal. 74(1): 204-216.

Heard, W.R. 1991. Life history of pink salmon. Pp. 119-230 in Pacific Salmon Life Histories, C. Groot and L. Margolis, eds. Vancouver, B.C.: UBC Press.

Kettlewell, H.B.D. 1961. The phenomenon of industrial melanism in Lepidoptera. Annu. Rev. Entomol. 6:245-262.

Kluge, A.G. 1990. Species as historical individuals. Biol. Philos. 5:417-431.

Lack, D. 1971. Ecological Isolation in Birds. Cambridge, Mass.: Harvard University Press.

Lancaster, J. 1990. Lujan: Endangered Species Act too tough, needs changes. Washington Post, May 12.

Margulis, L., and K.V. Schwartz. 1987. Five Kingdoms: An Illustrated Guide to the Phyla of Life on Earth. 2nd Ed. New York: W.H. Freeman.

Mayr, E. 1942. Systematics and the Origin of Species from the Viewpoint of a Zoologist. New York: Columbia University Press.

Mayr, E. 1963. Animal Species and Evolution. Cambridge, Mass.: Belknap.

Mayr, E. 1969. Principles of Systematic Zoology. New York: McGraw-Hill.

Mayr, E., and P.D. Ashlock. 1991. Principles of Systematic Zoology. 2nd Ed. New York: McGraw-Hill.

Mishler, B.D. 1985. The morphological, developmental, and phylogenetic bases of species concepts in bryophytes. Bryologist 88:207-214.

Mishler, B., and R. Brandon. 1987. Individuality, pluralism, and the phylogenetic species concept. Biol. Philos. 2:397-414.

Mishler, B.D., and M.J. Donoghue. 1982. Species concepts: A case for pluralism. Syst. Zool. 31:491-503.

Nei, M. 1987. Molecular Evolutionary Genetics. New York: Columbia University Press.

Nixon, K.C., and Q.D.Wheeler. 1990. An amplification of the phylogenetic species concept. Cladistics 6:211-223.

NOAA (National Oceanic and Atmospheric Administration). 1991. Policy on applying the definition of species under the Endangered Species Act to Pacific salmon. Fed. Reg. 56:58612-58618.

NRC (National Research Council). 1995. Upstream: Salmon and Society in the Pacific Northwest. Washington, D.C.: National Academy Press.

O'Hara, R.J. 1993. Systematic generalization, historical fate, and the species problem. Syst. Biol. 42:231-246.

Otte, D., and J. Endler. 1989. Speciation and Its Consequences. Sunderland, Mass.: Sinauer Associates.

Parsons, T.J., S.L. Olson, and M.J. Braun. 1993. Unidirectional spread of secondary sexual plumage traits across an avian hybrid zone. Science 260:1643-1646.

Paterson, H.E.H. 1985. The recognition concept of species. Pp. 21-29 in Species and Speciation, E.S. Vrba, ed. Transvaal Museum Monogr. 4. Pretoria: Transvaal Museum.

Paterson, H.E.H. 1993. Evolution and the recognition concept of species: Collected writings, S.F. McEvey, ed. Baltimore, Md.: The Johns Hopkins University Press.

Peterson, A.T. 1992. Phylogeny and rates of molecular evolution of the *Aphelocoma* jays (Corvidae). Auk 109:133-147.

Raven, P.H. 1974. Plant systematics 1947-1972. Ann. Mo. Bot. Gard. 61:166-178.

Richards, R.L. 1971. The woodrat in Indiana: Recent fossils. Proc. Indiana Acad. Sci. 81:370-375.

Ridley, M. 1989. The cladistic solution to the species problem. Biol. Philos. 4:1-16.

Roelke, M.E., J.S. Martenson, and S.J. O'Brien. 1993. The consequences of demographic reduction and genetic depletion in the endangered Florida panther. Curr. Biol. 3:340-350.

Smith, C.R., D.M. Pence, and R.J. O'Connor. 1993. Status of neotropical migratory birds in the Northeast: A preliminary assessment. Pp. 172-188 in Status and Management of Neotropical Migraptory Birds, D.M. Finch and P.W. Stangel. eds. Gen. Tech. Rep. RM-229. U.S. Department of Agriculture Forest Service, Rocky Mountain Forest and Range Experimental Station, Fort Collins, Colo.

Solbrig, O. 1970. Principles and Methods of Plant Biosystematics. Toronto: Macmillan.

Steadman, D.W., L.J. Craig, and J. Bopp. 1993a. Diddly Cave: A new late Quaternary vertebrate fauna from New York State. Curr. Res. Pleist. 9:110-112.

Steadman, D.W., L.J. Craig, and T. Engel. 1993b. Late Pleistocene and Holocene vertebrates from Joralemon's (Fish Club) Cave, Albany County, New York. Bull. N.Y. State Archaeol. Assoc. 105:9-15.

Stebbins, G.L. 1950. Variation and Evolution in Plants. New York: Columbia University Press.

Stebbins, G.L. 1970. Biosystematics: An avenue toward understanding evolution. Taxon 19:205-214.

Stuessy, T. 1990. Plant Taxonomy. New York: Columbia University Press.

Templeton, A. 1989. The Meaning of Species and Speciation: A Genetic Perspective. Pp. 3-27 in Speciation and Its Consequences, D. Otte and J. Endler, eds. Sunderland, Mass.: Sinauer Associates.

U.S. Congress. 1975. Committee on Merchant Marine and Fisheries 94-A, House Document 94-51. Washington, D.C.: U.S. Government Printing Office.

Waples, R.S. 1991. Pacific salmon, *Oncorhynchus* spp., and the definition of "species" under the Endangered Species Act. Marine Fish. Rev. 53:11-22.

Waples, R.S. In press. Evolutionarily significant units and the conservation of biological diversity under the Endangered Species Act. In Evolution and the Aquatic Ecosystem, J.L. Nielsen and D.A. Powers, eds. American Fisheries Society, Bethesda, Md.

Weir, B. S. 1990. Genetic Data Analysis. Sunderland, Mass.: Sinauer Associates.

Wilcove, D.S., M. McMillan, and K.C. Winston. 1993. What exactly is an endangered species? An analysis of the U.S. Endangered Species list: 1985-1991. Conserv. Biol. 7(1):87-93.

Wiley, E. 1981. Phylogenetics: The Theory and Practice of Phylogenetic Systematics. New York: John Wiley & Sons.

Wiley, R.W. 1980. *Neotoma floridana*. Mamm. Species 139:1-7.

Woolfenden, G.E. and J.W. Fitzpatrick. 1984. The Florida Scrub Jay: Demography of a Cooperative Breeding Bird. Princeton, N.J.: Princeton University Press.

4

The Role of Habitat Conservation and Recovery Planning

THE IMPORTANCE OF HABITAT

Habitat is the physical and biological setting in which organisms live and in which the other components of the environment are encountered (Krebs, 1985; Jones, 1987). The concept of habitat is critical to modern ecology and was adopted and promulgated in some of the earliest treatises and texts (Elton, 1927; Clements and Shelford, 1939); habitat is a basic requirement of all living organisms (e.g., McNaughton, 1989). Habitat is one of the four components of a species' environment, along with climate variables, nutrients, and other interacting organisms (Andrewartha and Birch, 1954). The fact that habitat serves a multitude of organisms is critical to understanding its full role in the Endangered Species Act. Many species have not been classified, nor their status determined. Our knowledge of species is too limited, and the species deserving of endangered or threatened status too numerous, to list all that might merit it in a time frame adequate to protect them. The number of unclassified living species is thought to be from two to ten times the number that have been identified and named (Wilson, 1992). Although this gap in knowledge is greatest in the tropics, invertebrates are also incompletely described in temperate habitats, particularly for soil organisms and many marine groups. Even among named species in the United States, genetic diversity of subspecies and of vertebrate population segments is poorly understood, except for relatively few vertebrates that have received special attention because they are commercially or recreationally important. Prospects are scant for a significant, near-term change in this information gap, because the task is so immense, and there are few specialists for many plant or invertebrate taxa. Therefore, broad, ecosystem-based conservation measures that depend neither on taxonomic knowledge nor on the determination of listing status of individual species are also needed to prevent species extinction (see Chapter 10).

Fortunately, the distribution of habitats on which species diversity depends is somewhat better understood. In California, for example, reliable inventories document the loss of habitat and the extent of many valuable remnants, including coastal and inland wetlands, native coastal dunes, coastal sage scrub, oak woodlands, vernal pools, and free-flowing rivers (Jensen et al., 1993). Many endangered species are also found in California, as the Endangered Species List (50 CFR 17.11 & 17.12) shows. The relationship, nationwide, between vanishing habitats and vanishing species is well documented (see Chapter 2).

Based on studies of the relationship between habitat area and several groups of organisms, a simple and fairly general ecological relationship has emerged: species diversity is positively correlated with habitat area (McArthur and Wilson, 1967; Pianka, 1978). The relationship is $S = CA^z$; where S is the number of species, A is the area of the habitat, and C is an empirically determined multiplier that varies from place to place and among taxa. The exponent z varies according to topographic diversity and isolation of the habitat; z is usually larger for islands (around 0.3) than for mainland habitats (often less than 0.2). A corollary of this relationship is that if habitat is substantially reduced in area or degraded, species will be lost. The loss is not linear; a loss of 90% of the habitat results in the estimated loss of 30%-60% of the dependent species (Groombridge, 1992). The species loss might be even greater than 30%-60% for at least three reasons. First, we lack sufficient taxonomic and ecological understanding of many taxa that might have demands on habitat area quite different from those of the more intensively studied taxa. Also, lack of knowledge about a taxon can lead to an incorrect evaluation of the species-area relationship for it, because it can lead to misidentification of species. Second, many species-area relations were developed for nonmigrant species, tropical species, and true island populations—conditions that strictly apply to few North American taxa. Third, habitat can be altered in many ways, and the effects of most of these alterations on species numbers have not been studied.

Despite uncertainties in the actual mathematical relationship between habitat size and the number of species in that habitat, there is no disagreement in the ecological literature about one fundamental relationship: sufficient loss of habitat will lead to species extinction (see Chapter 2). And habitat protection is a prerequisite for conservation of biological diversity. Habitat protection is essential not only to protect those relatively few species whose endangerment is established, it is also in essence a pre-emptive approach to species conservation that can help to avoid triggering the provisions of the Endangered Species Act.

Loss of habitat is a major factor in species extinctions when the cause of extinction is known. Groombridge (1992) provided a table showing that for animals with known causes of extinction, hunting (mostly unregulated) caused extirpation of 23%, introduced animals caused 39%, and habitat loss accounted for the loss of 36%. Habitat loss has been more important as a recent cause of extinction of terrestrial than marine species to date and probably accounts for more than 36% of the extinctions where the cause is unknown, because extirpation due to unregulated hunting is easily documented.

Habitat loss places additional pressures on endangered species management because where the amount of habitat available is limited, protection or maintenance of the present condition of a habitat for one species can adversely affect another (see Chapter 6). For example, boundaries or edges between different habitat types (gaps or openings) favor some wildflowers and other desirable plants (Whitford, 1949; Gilbert, 1980), but large, unbroken tracts of habitat favor other species. When the amount of habitat available is limited, the design of optimal combinations of mosaics of distinct habitats across landscapes becomes a major challenge to managers.

THE ROLE OF HABITAT CONSERVATION UNDER THE ESA

The role of conserving habitat for endangered species has been recognized since the first federal endangered species legislation. For example, the Endangered Species Preservation Act of 1966 (P.L. 89-699), stated, "It is . . . the policy of Congress that the Secretary of Interior, the Secretary of Agriculture, and the Secretary of Defense . . . shall preserve the habitats of such threatened species on lands under their jurisdiction" (Section 1(b)). Over time, as our knowledge of species requirements has grown, the legislation has evolved from the regulation of harvest and trade in species to the protection of habitat. The stated purposes of the current ESA are to conserve endangered species "and the ecosystems on which they depend" (16 U.S.C. 1531), a clear mandate linking successful conservation of species to the habitats that they require. This linkage is entirely appropriate scientifically.

The ESA provides throughout for the identification and protection of habitat. The first statutory consideration for the *listing* of species as threatened or endangered is "the present or threatened destruction, modification or curtailment of its habitat or range" (§1533). Section 4 of the act further requires the designation of a species' "critical habitat" concurrently with the listing of a species, unless earlier listing is "essential to the conservation" of the species (§1533(b)(6)(C)(8)) or unless the designation of critical habitat is not "prudent" or "determinable" (§ 1533(a)(3)). Critical habitat designations are, "to the maximum extent practicable," to be accompanied by "a brief description and evaluation of those activities (whether private or public) which, in the opinion of the Secretary, if undertaken may adversely modify such habitat, or may be affected by such designation" (§1533(b)(6). Critical habitat is to be designated "on the basis of the best scientific data available and after taking into consideration the economic impact, and any other relevant impact, of specifying any particular area as critical habitat" (§1533(a)(3)). (Critical habitat designation is not mandated for the species listed before 1978, when these provisions were added. To date, critical habitat has been designated for less than 20% of all species listed in the United States. Furthermore, the ESA provides for exclusion of areas from critical habitat if it is determined that "the benefits of such exclusion outweigh the benefits of specifying such area," unless failure to designate the area "will result in the extinction of the species concerned" (§1533(b)(2)).

Section 7 requires federal agencies to *consult* with the Fish and Wildlife Service (FWS) and the National Marine Fisheries Service (NMFS) to ensure that federal actions are not likely to jeopardize the continued existence of listed species or result in "the destruction or adverse modification" of their critical habitat (§1536). Habitat modification gives rise to the vast majority of these consultations, nearly all of which are resolved informally or with modifications that allow projects to proceed as planned.

Sections 9 and 10 of the act have extended the review of habitat modification to nonfederal, private development. (However, recent court decisions have left uncertain the validity of FWS regulations that state that habitat modification constitutes "take.") Section 9 prohibits "take" of a listed species, a term described elsewhere in the act as, among other things, "harm" to a species. Early FWS regulations described this harm to include "significant habitat modification or degradation" that "kills or injures wildlife" (50 CFR 17.3[1])). The act's application to habitat modification by private parties led to the development of early conservation plans in California, a process endorsed by the Congress in 1982 with the additions to Section 10. Section 10 (a) currently requires "habitat conservation plans"

---

[1]The Supreme Court recently reviewed *Babbitt versus Sweet Home Chapter of Communities for a Greater Oregon*. The case focused on FWS regulations concerning the act's prohibition against taking an endangered species, in this case, the northern spotted owl. *Take* is defined in the ESA to include *harm*, and FWS has defined *harm* to include significant habitat modification that adversely affects an endangered species. The U.S. Court of Appeals for the District of Columbia ruled in a split decision that "the Service's definition of 'harm' was not clearly authorized by Congress nor a 'reasonable interpretation' of the statute" (1994 U.S. App. LEXIS 4341); the Supreme Court decided 6-3 that Secretary Babbitt reasonably construed Congress' intent when he defined "harm" to include habitat modification (115 S. Ct. 2407).

of private parties seeking to secure an "incidental take" permit (unintentional take) for listed species (§1539(a)(2)).

One exception to the habitat conservation planning requirements of Sections 9 and 10 is worth noting. Although Section 7 constrains federal agencies from jeopardizing plant and animal species, no such constraints are imposed on private parties for the incidental taking of listed plants. Private parties are forbidden only from "removing" or "maliciously" damaging or destroying endangered plants found either on federal lands or under the protection of state law (§1538(a)(2)). However, plants are covered by habitat conservation plans in that the Section 10(a) permit requires a consultation under the Section 7 requirements.

Recovery planning under Section 4 of the act is likewise keyed to habitat protection. Each recovery plan is to include "site-specific management actions as may be necessary to achieve the plan's goal for the conservation and survival of the species" (§1533(f)(1)(B)). Because most species are endangered due to loss or degradation of habitat, site-specific actions should include identification, restoration, and management of habitat.

Habitat acquisition for endangered species has also been a part of the federal program from its beginnings. Later amendments to the ESA have augmented the authority and funding for this effort, but acquisition has not kept and cannot keep pace with the number and size of the affected habitats or the modification and degradation that they face.

In summary, habitat protection has always been an important component of endangered species programs. As our experience with endangerment and recovery has increased, habitat has become the central ingredient, and the ESA, in emphasizing habitat, reflects the current understanding of the crucial biological role habitat plays for species.

CRITICAL HABITAT AND FEDERAL ACTIVITIES

Section 7 imposes a special requirement on federal agencies to ensure that their activities are "not likely to . . . result in the destruction or adverse modification of habitat of [listed] species which is determined by the Secretary, after consulting as appropriate with affected States, to be critical . . .". Critical habitat is a valid biological concept. The requirement of Section 7 corresponds to the understanding of conservation biology that certain habitat is essential for species survival. Habitat critical to species can be identified from the knowledge of species and ecosystems as objectively and scientifically as a species can be identified for listing. However, as is the case with listing decisions on many rare species, detailed information needed to designate critical habitat might be lacking. Simple occurrence of a species within a habitat does not necessarily mean that the habitat is required by the species (Van Horne, 1983) or that the amount and quality of habitat might be considered "critical." But that a species is absent from a given habitat does not mean that the habitat is not critical to the persistence of the species (see Chapter 5). Identification of the relationship of a species to habitat and the determination of what is critical to the long-term survival of that species are high priorities for long-term conservation. The complexity of designating critical habitat will vary by species, but designation should be possible in many cases. Mechanisms exist to uncouple critical habitat designations from listing when they need additional investigation and would otherwise cause delay. Mechanisms also exist to withhold these designations where they are not "determinable" or otherwise "prudent." That nearly 80% of all species listed do not have critical habitat designations is a cause for concern.

Survival Habitat

The question has been raised whether critical habitat should be determined at the time of listing or whether it should be deferred to the time of recovery planning. The advantages of early designation include the provision of some "early warning" to all parties, and in particular, the affected federal agencies, that such areas are to be treated with particular caution. Designated habitat is protected by a more objective standard ("no adverse modification") than that provided for threats to species ("no likelihood of jeopardy") in that adverse habitat modifications are more amenable to objective measurement and quantification than are the many factors that might contribute to jeopardizing the survival of a species. The standard of habitat protection provides an important point of focus for those outside of government, including the scientific community, to help protect areas at least until recovery plans are developed that will clarify the needs of endangered species and provide more fully for their recovery. Importantly, critical habitat designation can be beneficial to other listed and nonlisted species living in the designated area, especially for those species for which satisfactory long-term recovery plans have not been implemented.

The committee recognizes that because of public concern over economic consequences, the designation of critical habitat is often controversial and arduous, delaying or preventing the protection it was intended to afford. Because critical habitat plays such an important biological role in endangered species survival, we believe that some core amount of essential habitat should be designated at the time of listing and should be identified without reference to economic impact. We recognize, however, that economic review may need to remain linked to critical habitat determination in the ESA, and that determination of areas essential to the recovery of a species, including areas not currently occupied by that species, can be especially complex. Hence, we recommend designation of survival habitat for endangered species as follows:

1. *Survival habitat* would be designated at the time of listing, unless insufficient information were available or harm to the species would occur. For this purpose, survival habitat would mean that habitat necessary to support either current populations of a species or populations that are necessary to ensure short-term (25-50 years) survival, whichever is larger.[2] Survival habitat would receive the full protection that the Endangered Species Act accords to critical habitat, and the adverse modification standard of Section 7 (a)(2) would apply. No economic evaluation would be conducted. The purpose of this requirement is to preserve scientific and management options until recovery plans are in place and effective.

2. The designation of survival habitat (and its protection under the ESA) would automatically expire with the adoption of a recovery plan and the formal designation of critical habitat. This underscores the emergency and temporary nature of survival habitat. It is intended to be a way of avoiding delays and providing immediate protection while a more comprehensive evaluation of habitat requirements is performed, not as a substitute for critical habitat designation. Expedited review and designation of survival habitat should be made for each currently listed species whose critical habitat has not yet been determined, unless recovery planning, including critical habitat designation, is expected within the period required for survival habitat designation.

3. Subsequent recovery planning would include designation of critical habitat as currently defined in the ESA (including economic evaluation) to include areas necessary for species recovery.

Because in our recommendation essential survival habitat is identified without reference to economic impact, and because it might not be sufficient to ensure long-term survival and recovery of endangered species, the committee views it as an emergency, stop-gap measure until critical habitat can be designated and a recovery plan can be completed, not as a substitute for those measures. To avoid harm that indefinite delays in designating critical habitat and formulating recovery plans might cause to economic interests and to the endangered species itself, the implementation of this recommendation needs to include ways of preventing those delays from occurring.

---

[2]We do not specify a precise time, because the length of survival time expected should be based on knowledge of the species' biology, including its generation time.

PRIVATE ACTIVITIES AND
HABITAT CONSERVATION PLANNING

Endangered species are found across North America, and slightly more than 59% of land in the United States is privately owned (about 8% is under state and local ownership and about 33% is federally owned (NRC, 1993)). Clearly, a program that targets only the portion of the United States that is under federal control will not prevent species extinction. Habitat conservation planning under Section 10 and, more recently, Section 4(d), has addressed species protection on private lands. Two examples of habitat conservation planning are described below.

Habitat-Conservation Plans

In Section 7 of the ESA, Congress offered relief from Section 9 prohibitions against taking endangered species where the government's actions were "not likely to jeopardize the continued existence of any endangered or threatened species." Similar relief was not available to nonfederal parties when federal permits were not required for development or other activities. In its 1982 amendments to the ESA, Congress provided nonfederal parties the possibility of obtaining legal exemption from Section 9 prohibitions against take of endangered species (and, by regulation, of threatened species). It amended the ESA to provide private nonfederal parties "incidental take" permits under Section 10(a), if a habitat conservation plan (HCP) is submitted that, when implemented, "will, to the maximum extent practicable, minimize and mitigate the impacts of such taking" and "not appreciably reduce the likelihood of survival and recovery of the species in the wild" (16 U.S.C. 1539(a)(2)(A)).

Congress also intended to provide a framework that would encourage creative partnerships between the private sector and local, state, and federal agencies in the protection of endangered species and habitat conservation. The legislators pointed to the model of the San Bruno Mountain HCP,[3] developed by private landowners and local, state, and federal government agencies to conserve the endangered mission blue butterfly (*Plebejus icarioides missionensis*) and several co-occurring species in coastal California. Although it focused on listed and candidate species, the HCP addressed the San Bruno Mountain ecosystem as a whole, including its diversity of species and habitats. The plan included protection of open space and habitat diversity; management, including control of exotic species and habitat restoration; protection of sensitive species during project construction; funding

---

[3]H.R. Rep. No. 835, 97th. Congress, 2nd session 31, reprinted in 1982 U.S. Code Congressional and Administrative News 2807, 2831.

of plan activities; and assurances to the private sector that new requirements would not be imposed after the plan was accepted (Thornton, 1991). Most of the grassland habitat of the butterfly was protected, and its population more than a decade later is as large as it was before the plan was implemented.

The model had its second test in southern California, where the threatened Coachella Valley fringe-toed lizard (*Uma inornata*) survived in remnant sand-dune habitats, at risk from suburban expansion around Palm Springs. Challenged to preserve both dune habitat and distant montane sources of sand, FWS ultimately agreed to a reserve system of nearly 17,000 acres. However, this habitat constitutes barely 10% of the occupiable habitat that existed at the time the HCP was initiated, and little is known even now about what characteristics of the lizard's biology cause its dramatic population fluctuations and put it at risk of extinction (Beatley, 1994).

These plans have had few imitators. Several promising regional HCPs have collapsed before completion. Two of the more notable ones are the North Key Largo (Florida) HCP, designed to conserve the American crocodile (*Crocodylus acustus*), the Key Largo woodrat (*Neotoma floridana smalli*), the Key Largo cottonmouse (*Peromyscus gossypinus allapaticola*), and other species; and, very recently, the Balcones Canyonlands HCP (near Austin, Texas), which focused on the black-capped vireo (*Vireo antricapillus*) and the golden-cheeked warbler (*Dendroica chrysoparia*), among other species and their habitats.

Although FWS provides detailed guidelines on administrative procedures for developing HCPs (FWS, 1990; 1994), its directives in the application of biological data to plan development are very sparse. The guidelines state that on request, FWS will indicate whether the biological data are adequate to proceed with the other elements of this process. If desired, FWS will recommend the number, type, scope, and general design of studies needed to provide acceptable data to develop the conservation plan. FWS, however, has had remarkably few opportunities to exercise this offer, because completed HCPs have been so few.[4]

Critics have argued that HCPs demand inordinate amounts of time, human resources, and money, and they should be avoided if possible (e.g., Mann and Plummer, 1995). Consultants encourage private landowners to use any federal "hook" (i.e., relationship to some federal activity) possible to avail themselves of the more expeditious Section 7 consultation process. Reasons for the limited number of completed HCPs are myriad, but at the

---

[4]From 1982 to July 1994, FWS issued 33 incidental take permits and 12 permit amendments, but many of those were very small projects rather than full-scale HCPs. Approximately 130 HCPs and permit applications are now in various stages of development (FWS, 1994), most of them small in scope as well.

very least, increased funding is necessary to upgrade the active role of FWS in the process. Furthermore, agency staff need the biological guidelines that FWS offers to those who seek permits under Section 10(a). Appropriate guidelines should be developed to assist planners in applying biological data to habitat-conservation planning. They should include much of the agenda discussed in this report under recovery planning, as well as elements from California's Natural Communities Conservation Plan described below. Finally, they should attempt to resolve dilemmas that might arise if the need for ecosystem protection were to conflict with the legal requirement of protection of individual species; in other words, an ecosystem should not fail to receive protection because of the needs of a single species. Wherever possible, HCPs should be regional in scope and should serve multiple species across multiple habitat types. Model programs are in process concerning the scrub community of Brevard County, Florida, and the southern San Joaquin Valley of southern California.

At a minimum, guidance should be offered in (1) development of explicit reserve design and management goals and objectives, (2) identification of techniques and data needed to perform population-viability analysis or equivalent demographic or metapopulation modeling efforts to assess likelihoods of persistence of target species under alternative planning options, (3) description of management options and discussion of how ongoing research and monitoring activities will be used to adjust management in response to changes in population sizes and environmental variables, (4) application of risk analyses in consideration of plan alternatives, and (5) description of how these exercises should be applied in the land-use planning process. Guidelines thus should describe a discrete scientific program for HCPs that will be specific, efficient, and cost-effective (although it will usually be impossible to make HCPs inexpensive).

RECOVERY PLANNING

The ultimate goal of the ESA is to recover threatened and endangered species. Recovery is described as "the process by which the decline of a threatened or endangered species is arrested or reversed, and threats to its survival are neutralized, so that its long-term survival in nature can be ensured" (FWS, 1992). The 1978 amendments to the act first required that, following species listing, a recovery plan be developed (§1533(f)). Faced with delays in this part of the planning process and with plans too vague to provide meaningful guidance, Congress has increased funding and provided more detailed requirements for the contents of these plans (§ 1533f(1)(B)). Despite this increased attention, recovery plans are developed too slowly. There appear to be legitimate concerns about the adequacy of recovery plan objectives and commitments to implement them. Plans that involve species of little public interest often sit unimplemented for long periods, especially if they would be expensive to implement (Tear et al., 1993).

The backlog in recovery planning is significant because it is large, and because nearly everything else in the act can be seen as a preliminary measure (e.g., listing) or as protecting the options for recovery plans (e.g., critical habitat designation or consultation). In 1988, more than 10 years after recovery plan requirements were first added to the statute, only 56% of listed species had recovery plans approved, with another 18% in preparation. In 1992, some 61% of species had approved plans, but by the following year the percentage had dropped to 53%, a figure attributable to recent accelerated listing actions. As of March 31, 1995, resource agencies had approved 411 plans covering 513 species, 54% of the 956 U.S. species listed at that time.

A FWS report to Congress (FWS, 1992)—the most recent data available to the committee—evaluated recovery progress as of September 30, 1992, based on the percentage of tasks accomplished for downlisting, delisting, or maintaining current populations for the foreseeable future. Managers had attained greater than 50% of their recovery objectives for only 68 species, while 544 species (77% of listed species, including all recently listed species) had less than 25% of their recovery tasks completed. However, the report noted that 201 species, 28% of the total, were stable and 69 species (10%) were improving.

A review of 314 recovery plans for threatened and endangered species as of 1991 (Tear et al., 1993) concluded that

- Only about 17% of the plans contained population-size data;
- Nearly one-third of the plans with population data set recovery goals at or below the population size believed to exist at the time of listing; and
- Sixty percent of the vertebrates would remain at imminent risk of extinction with about a 20% probability of extinction within 20 years or 10 generations, whichever is longer, even if every population target were met.

In the recovery planning guidelines, recovery planners are asked to adhere to an explicit format that includes (1) an introduction that describes pertinent ecological, genetic, and other information related to the biology of the listed species and threats to it; (2) a recovery objective and criteria to meet the objective; (3) an implementation schedule, including priorities of tasks and cost estimates; and (4) an appendix identifying appropriate reviews of the plan and peripheral but pertinent documents or communications. Recovery plan tasks 1 and 4 are by and large straightforward; unfortunately, tasks 2 and 3 are not, for different reasons.

The guidelines correctly observe that the quantification of "recovery criteria calls for creative thought and developing the criteria may require educated guesswork." Nonetheless, they demand "concise and measurable recovery criteria" that will serve as "the central pillar of the recovery plan." Population viability analysis is the cornerstone, the obligatory tool by which recovery objectives and criteria are identified. Yet the demographic and genetic data necessary to fuel such analyses are lacking for virtually all species for which recovery plans exist, and likewise for those in and awaiting entry into the recovery-planning process.

In the absence of adequate data, "educated guesswork" has and will continue to rule the development of the "central pillar" of recovery plans. Moreover, few scientists agree on the data and analyses that are required to produce a reliable population viability analysis for conservation planning purposes. What most will agree is that many obligatory elements, such as variance in reproductive success, are difficult to quantify and that the time-series data necessary to reduce the confidence intervals around projections from population viability analyses demand long-term and expensive field research. In essence, even to set objectives and criteria for recovery in a scientifically defensible manner (much less to carry them out) will demand resources well beyond those currently available. Nevertheless, many analytical tools that were unavailable to planners in the past now could be used to add greater scientific rigor to recovery plans (see also Chapter 7). However, the recovery guidelines offer no explicit schedule or outline for bringing data to bear in setting recovery objectives or criteria. The committee believes that FWS should convene a working group to develop explicit guidelines for the application of data to the construction of recovery objectives and criteria, with particular emphasis on the estimation of risks and the rational establishment of risk criteria.

Planners would receive general guidance covering at least the following topics:

1. A habitat-based, *in situ* approach to recovery that puts *ex situ* actions (e.g., captive rearing) in their appropriate context.

2. A logical hierarchical approach to the analysis of available data on the species and its habitats and to the acquisition of additional information on the species of concern and the habitats that support it, including predictions from ecological theory and population dynamics models and inferences drawn from related species.

3. Guidance for the application of population viability analysis in conservation planning that clearly identifies in planning prescriptions the inherent uncertainty that accompanies such demographic modeling exercises.

4. An outline that describes in detail future research needs and how that research when completed will contribute to species and habitat management.

5. An effective monitoring scheme that relates census information to the physical and biotic factors likely to affect population dynamics.

Because prioritizing activities is necessary but difficult, and because estimating cost would be helpful but nearly impossible for typical recovery teams to carry out, the committee recommends that the prioritization and estimation of costs currently called for in recovery planning be reviewed.

Recovery planning remains handicapped by delays in implementation, the scientific validity of objectives, and the uncertainty of application to other federal activities. To be scientifically credible and to increase their likelihood of being successful, recovery plans should consider at least the following questions.

How much of a species' historic range should be protected to ensure recovery or prevent extinction?

Are there critical aspects of a species' life history or ecological and genetic requirements that must be known to successfully implement a recovery plan? Are they known?

What is known about the species' use of and need for corridors among its various populations?

To what degree is a focus on a single species likely to be as successful as an approach that includes the needs of other species in the area, or an ecosystem approach?

A final desired feature is that recovery plans reflecting the best judgment of science from decisions made before listing through consultation and habitat planning bear some relationship to decisions affecting the future of the species. We recommend, therefore, that all recovery planning include an element of "recovery plan guidance," particularly with regard to activities anticipated to be reviewed under sections 7, 9, and 10 of the ESA. To the extent feasible, the guidance should identify activities that can be assumed to be consistent with the requirements of those sections, activities that can be assumed to be inconsistent with them, and activities that require case-by-case evaluation. The guidance should also specify criteria for use in preparing habitat management plans by persons seeking authorization for activities under Section 10 of the ESA and for the planning of federal agencies in furtherance of their Section 7 responsibilities. These measures are fundamental to many recovery efforts.

The real issue is that no recovery plan, however good it might be, will help prevent extinction or promote recovery if it is not implemented expeditiously. Indeed, the failure to implement a recovery plan quickly can also increase the disruption of human activities, through uncertainty, among other causes.

Funding will be required to develop sound recovery plan guidance and the recovery plans themselves. Unfortunately, recovery planning under the

ESA has a history of large ideas and little follow-through. Expenditures for recovery have increased in recent years but remain meager compared with the billions of dollars likely to be needed to attain recovery goals for the bulk of listed species (DOI Office of the Inspector General, 1990; Jackson, 1992). The FWS appropriation to support recovery programs was $10.6 million in FY 1990. For FY 1995, it was $39.7 million. More realistic budgets are essential. Mechanisms should be considered to facilitate funding of complex plans, including special funds for plan implementation and funding agreements with affected parties (e.g., private, state, local, tribal, and other agencies).

We also suggest consideration of a new, comprehensive and continuing funding mechanism for developing and implementing recovery programs for listed species. General wildlife and migratory waterfowl programs at the federal and state level have been funded for decades through trusts based on hunting and fishing licenses and the sales of arms and ammunition. Federal restoration programs for hazardous waste sites, abandoned surface mines, and the nation's coastal zone are similarly funded by dedicated revenue sources. Some state acquisition and restoration programs are likewise based on user fees and on real-estate transfers. These earmarked funding sources are widely accepted and successful in providing stable revenue on which long-term planning decisions can be based. Planned programs to recover listed species would benefit enormously from similar support.

The following description of the Natural Communities Conservation Program for the coastal sage-scrub communities of southern California is an example of how many of the recommendations that the committee has provided are being implemented in a program that could serve as a model for better recovery planning and permitting under sections 7 and 10(a).

NATURAL COMMUNITY CONSERVATION PROGRAM AND COASTAL SAGE SCRUB COMMUNITY OF SOUTHERN CALIFORNIA

Recent listings of the northern spotted owl, desert tortoise, fish species from the Sacramento-San Joaquin delta estuary, and California gnatcatcher have affected economic activity in several regions of California. Secretary of the Interior Bruce Babbitt has advocated new and innovative ways to avoid these "economic train wrecks," favoring approaches that take a comprehensive view of endangered species protection. His strategy for listing the California gnatcatcher by using a special rule under Section 4(d) of the ESA in cooperation with California's Natural Community Conservation Plan (NCCP) allows for moderate economic growth while conservation plans for the gnatcatcher are developed that also consider other unlisted species in the coastal sage scrub.

The goal of NCCP is to conserve healthy functioning ecosystems and the species that are supported by them. The pilot program for this process focused on the coastal sage scrub ecosystem in southern California. Remnant, coastal sage scrub habitats support a diverse assemblage of native animals and plants, including the imperiled California gnatcatcher, coastal cactus wren, more than 20 other vertebrate species that are candidates for federal protection, several invertebrate candidate species, as well as nearly 100 plant species that are of special conservation concern.

The first NCCP was initiated with appointment of a scientific review panel enjoined to outline a conservation strategy for the coastal sage scrub ecosystem. Although the scientific review panel collected extensive information on the coastal sage scrub community and its constituent plant and animal species, it was not able to produce final scientifically defensible guidelines for long-range planning purposes because the available database was limited. Instead, the scientific review panel assisted in producing interim planning rules that institutionalize several of the recommendations in this chapter, including the interim protection of habitat crucial for survival until recovery planning is in place.

The scientific review panel concluded that federal protection of the California gnatcatcher alone would not be sufficient to confer protection to other species at risk in the coastal sage scrub community. Indeed, it is highly likely that prohibition of take of the gnatcatcher under Section 9 of the Endangered Species Act would not even serve to ensure the persistence of that species. Its highly disjunct distribution demands conservation of ample amounts of coastal sage scrub habitat, including habitat that is not currently occupied by the gnatcatcher, as well as conservation of other habitat types to provide landscape corridors for dispersal by the species. In the absence of a regional, community-level conservation plan, additional listings of coastal sage-scrub species would be inevitable despite the listing of the California gnatcatcher.

Toward the development of a community-level conservation plan and focusing on three target species of vertebrates (the California gnatcatcher, the coastal cactus wren, and the orange-throated whiptail lizard), the distributions of which embrace the majority of the planning area, the scientific review panel examined existing data and suggested additional information to be gathered that would shed light on the coastal sage scrub community and the many species it supports. This multiple-species planning effort attempted to use the best scientific methods available to:

1. Define the management planning region;

2.   Identify subregional planning areas that consider landscape features, biological factors, and political jurisdictions;

3.   Describe an interim conservation planning framework for developing a viable system of subregional reserves;

4.   Recommend appropriate techniques for gathering baseline field data and their analysis;

5.   Recommend research and monitoring activities to assist in subregional conservation planning for individual areas; and

6.   Recommend a selection process for a subcommittee of specialists to report on the best available techniques for restoration and management.

Some data available in some portions of the planning area were of substantial immediate value to subregional planning, but the amounts of information then available by and large were not adequate to identify the physical characteristics of management areas necessary to protect ecosystem functioning through time, to identify minimum viable population sizes for target species, or to describe effective landscape corridors that would facilitate ecological interaction and gene flow among organisms that occupy the coastal sage scrub community. The scientific review panel therefore described a research program focusing on six interactive research tasks.

1. Biogeography and inventory of coastal sage scrub. The basic extent and distribution of coastal sage scrub vegetation and its constituent species are to be adequately mapped for the region and for each subregion.

2. Trends in biodiversity. Monitoring of indicator taxa (such as coastal sage shrub dependent birds, small mammals, and butterflies) will help to assess the ongoing success of coastal sage shrub community conservation efforts. Relationships between species richness/composition and habitat patch area, and the effects of isolation are to be investigated in sampling programs. These sampling programs must include surveys for species richness and composition within a carefully selected series of coastal sage shrub patches in each subregion.

3. Dispersal characteristics and landscape corridor use. Data from several locations within the planning region are to be gathered during both breeding and nonbreeding seasons on target species, top predators, coyotes, and representative small mammals and invertebrates.

4. Demography and population viability analysis. Time-series data on the two target species of birds are to be gathered in at least half the subregions and from representative physical circumstances that span those found across the regional distributions of the species. Data must include territory size, time budgets, reproductive success, survivorship, emigration and immigration, with separate data obtained both for males and females where possible. Population viability analyses are to be carried out for sample

populations and metapopulations, and should consider connectivity and environmental effects.

5. Surveys and autecological studies of sensitive animals and plants. Basic information on the location, abundance, distribution, and natural history of vertebrate and invertebrate candidate species for federal protection and coastal sage shrub associated plant species of special concern are to be gathered from select sites throughout the planning region.

6. Genetic studies. The maintenance of genetic variation is critical to the long-term viability of species inhabiting coastal sage shrub and its assessment is an important aspect of population monitoring.

The NCCP planning area is in lands that have been greatly affected by human activities. Natural and anthropogenic disturbances will continue; many of those disturbances will reduce the capacity of coastal sage scrub habitats to support many species of concern. Areas designated as reserves thus are unlikely to be self-sustaining (that is, provide for natural, dynamic ecosystem processes) or to be capable of maintaining viable populations of target species without active management. The ability of individual patches of coastal sage scrub habitat to be effectively managed over the long term will be a critical factor in prioritizing of conservation efforts.

For these reasons and in recognition of the critical role of restoration activities in the Natural Communities Conservation Planning Program, the scientific review panel encouraged the immediate creation of a committee to address central issues in management and restoration to focus on

1.   Exotic species control, including animals and plants.

2.   Recreational use of coastal sage scrub and other open space reserve areas, including identification of suitable low impact recreational pursuits consistent with preservation goals.

3.   The role of fire in natural ecosystem dynamics and processes, including the application of managed burns and the control of ignitions of accidental and vandal origin.

4.   Identification of restoration unit sizes, including identification of maximum areas that are restorable using current techniques. A focus on patch enlargement techniques is advised.

5.   Identification of coastal sage scrub responses to soil conditions in restoration efforts, with focus on soil structure, soil nutrient levels, organic matter content, water holding capacity, and soil compaction.

6.   Identification of appropriate seeding, outplanting, and irrigation techniques with focuses on proper mixes of species, seeding techniques, and timing of applications of seed and irrigation.

7.   Identification of techniques to encourage native herbaceous species and to discourage the establishment of exotic species.

8. Establishment of realistic success criteria to evaluate management and restoration efforts considering sage species diversity and cover, and use by target species.

During the interim period while data are being gathered, land-use planning will continue. Draft conservation planning guidelines therefore were published that describe short-term land use restrictions for a limited period based on recommendations from the scientific review panel. Those recommendations were meant to allow some development without foreclosing future conservation options, and three key objectives are to be met during the interim data acquisition period.

First, development is to be limited to 5% or less of the landscape that is occupied by coastal sage scrub and its resident species. Development should also strive not to disproportionally affect any "environmental subunit" (defined by vegetational subcommunity, elevation, slope, aspect, latitude, distance from coast, and substrate) within each subregional NCCP planning area.

Second, to the extent feasible, development is to avoid likely "hotspots" of biotic diversity (based on habitat patch size and isolation).

Third, development is not to sever extant open space landscape linkages between biodiversity hotspots.

Because threats to the persistence of species were viewed as so pressing, conservation planning should result in no net loss of habitat value, defined as the ability of the coastal sage scrub habitat to support target species in a subregion over the long term. Expected incremental losses of habitat therefore must be mitigated with habitat restoration activities and effective management planning.

In addition, the scientific review panel believed that long-term NCCP goals were to be best met where six tenets of conservation biology are incorporated at all spatial scales in regional and subregional planning.

1. Species that are well distributed across their native ranges are less susceptible to extinction than are species confined to small portions of their ranges.

2. Large blocks of habitat containing large populations of the target species are superior to small blocks of habitat containing small populations.

3. Blocks of habitat that are close together are better than blocks far apart.

4. Habitat that occurs in blocks that are less fragmented internally is preferable to habitat that is internally fragmented.

5. Interconnected blocks of habitat serve conservation purposes better than isolated blocks, and habitat corridors or linkages function better when the habitat within them resembles habitat that is preferred by target species.

6.   Blocks of habitat that are roadless or otherwise inaccessible to humans serve to better conserve many target species than do roaded and accessible habitat blocks.

To differentiate between habitat areas that are likely to be more rather than less important to long-term planning, a habitat-evaluation process was developed.

Habitats are to be differentiated into three categories. Those having higher potential conservation value need to be identified early in the planning process and protected from habitat loss and fragmentation while planning is under way. The methodology described in the guidelines places 50% of the coastal sage scrub in each planning subregion in the higher potential value category. Habitats with intermediate potential value are those lands that probably cannot be managed as independent reserves, but which by virtue of high quality or proximity or linkage to higher value habitats should be treated as potentially significant for subregional conservation planning. Habitats of lower potential value are those left after the higher and intermediate value areas have been identified and include small, isolated patches, the loss of which would probably not affect the long-term viability of the target species or other species of concern. Overall, an estimated 10%-25% of the coastal sage shrub in a subregion would fall into the lower potential value category. For the ranking approach to interim habitat loss to function, it is important that a significant amount of land be classed as lower value. The criteria for identifying higher and intermediate value land are to be adapted to local conditions.

Each planning subregion needs to show interim protection for higher potential value lands on a map, using a step-down process. Large, dense areas of coastal sage scrub are the higher potential value lands. Natural lands that occur in linkages, that are close to possible higher value areas, or that have high species richness are considered intermediate potential value lands. Remaining coastal sage scrub is considered to have lower potential value. A guideline policy for local government treatment of the higher, intermediate, and lower potential value lands during the interim period includes six assessment criteria that address (in order) the presence of natural vegetation, the presence of coastal sage scrub, the size of the scrub patch, its proximity to higher value lands, its role as a potential landscape linkage or corridor, and the densities of target species.

HABITAT-RELATED STANDARDS

Several habitat-related features of the ESA differ without basis in science. Prominent among these are standards applicable to the protection of plants, and to the determination of jeopardy and modification of critical habitat. As noted earlier, Section 9 fails to protect endangered plants from habitat modification to the same degree that it protects animals. Plants, like animals, require places to live and space for dispersal of their off-spring. Plants include, for example, species isolated in unique environ-ments and others that disperse pollen, spores, and seeds over wide areas. The biological differences between animals and plants underlying their taxonomic separation offer no scientific reason for lesser protection of plants, nor do their importance to scientific and other research.

Under sections 7, 9, and 10, FWS and NMFS are required to determine whether federal actions or private actions that are federally permitted (Sec-tion 7) and nonfederal activities (sections 9 and 10) present an unacceptable threat to listed species and their habitats. Although the administrative pro-cesses for making these determinations differ, these judgments are not dis-tinguishable as matters of biology and physics. Two scientific issues need very careful attention in the application of the Endangered Species Act. The first concerns the differences and similarities between standards for public and private lands and actions. The second concerns the relationship between survival and recovery.

Public and Private Lands and Actions

Section 7 requires that federal and federally permitted actions do not jeopardize listed species; Section 9 requires that private parties not take them. The biological and physical requirements of species—including en-dangered and threatened species—do not vary according to the ownership of the habitats that they occupy. The requirement of sockeye salmon for clean, cool water and of clapper rails for safe nesting sites are not changed if they live on public or private land. A tidal marsh that provides habitat for endangered species will be equally affected if it is filled by a public agency for a public facility, such as a sports complex or a marina, or by a private agency for a private facility, such as a shopping center or a hotel. Therefore, there is no biological reason to have different standards for de-termination of "jeopardy," "survival," or "recovery" on public and on pri-vate lands.

This observation becomes particularly relevant to resolution of the con-troversy surrounding the application of sections 9 and 10 to habitat modifi-cation on nonfederal lands. Whether Congress intended sections 9 and 10 to regulate habitat modification on private lands is a legal question, but there is no scientific question that a species might be as effectively jeopar-dized by habitat development as by overharvest and that an endangered species program regulating habitat modification only on federal lands will fail to protect many endangered species. The committee also recognizes that public agencies and individual public servants on public lands behave differently from private landowners, both corporate and individual, on private lands, because their rewards, incentives, and disincentives are different.[5] Therefore, requirements applied equally on private and public lands will not necessarily provide the same degree of protection, although the biological standards or criteria on which the regulations are based are the same. It follows, then, that different mechanisms for avoiding endangerment and achieving recovery on public and private lands are probably needed.

Survival and Recovery

The act and its regulations distinguish between species survival and recovery to determine jeopardy to species and adverse modification of their critical habitat. Both determinations present biologists with judgments that differ in degree, not in kind. Survival and recovery are points on a continuum. Anything that jeopardizes species recovery makes survival less likely, although it might still be likely over the short term; to make survival very likely, recovery must be ensured. Ensuring survival for a short period is not equivalent to promoting recovery, especially for long-lived species like razorback chub in Lake Mead, where populations can persist for many years without reproduction. The difficulty arises when the concept of short-term survival is applied to determinations of jeopardy and of effect on critical habitat with respect to federal actions, such as development of water resources, which often have long-term, if not irreversible, effects. A decision, for example, to allow destruction of one of three remaining habitats for a species—predicated on the assumption that it will jeopardize only recovery and not survival—requires biologists to ignore threats to long-term survival by irreversible actions. The impacts of activities with long-term or irreversible consequences should be evaluated under a long-term standard. The relationship between survival and recovery should be kept clearly in the minds of those who make and implement decisions. It is important to make estimates of the probability of survival for given periods to help keep that relationship in mind.

CONCLUSIONS AND RECOMMENDATIONS

Habitat conservation and recovery planning are essential components of any program to protect endangered species. To prevent delay and to allow

---

[5]For example, Mann and Plummer (1995) described the destruction by private landowners of habitat on their land that could support endangered species so that they can avoid the prohibitions of Section 9 or avoid the need for a Section 10(a) permit for incidental take.

immediate action to protect endangered species, survival habitat should be designated for a period necessary to develop a recovery plan.

There is no biological or physical reason that standards relating to habitat protection, survival, and recovery should differ for plants and animals and for public and private lands. The degree to which public and private entities should bear the responsibilities of the Endangered Species Act is a policy decision. But there is no escaping the scientific conclusions that all species have certain biological and physical requirements no matter who owns the habitats, and that public and private landowners do not always respond in the same way to laws, regulations, and other incentives and disincentives.

Finally, survival and recovery are not equivalent standards, although they are related. Clearly, if a species does not survive, it cannot recover. It is less obvious, but still true, that any action that jeopardizes recovery also decreases the probability of long-term survival. To permit a rational evaluation of survival and recovery goals, it is important to provide estimates of probabilities of achieving various goals over various periods, and the periods should be expressed both in years and in generation times of the organism of concern.[6] It is also important that evaluation of long-term and irreversible impacts be conducted in terms of long-term recovery of the species.

The committee recommends that the impacts of activities with long-term or irreversible consequences be evaluated under a long-term standard.

The committee further recommends that the relationship between survival and recovery be kept clearly in the minds of those who make and implement regulations. To further that end, the committee recommends that determinations of survival and recovery contain good-faith estimates of the probability of survival for given periods, perhaps 5, 10, and 100 years, as well as the probability of survival in *numbers of generations*, perhaps 5, 10, and 100.

The committee endorses regionally based, negotiated approaches to the development of habitat conservation plans. Guidance from FWS for the development of such plans should include advice on the development of biological data, such as demographic and genetic analyses, habitat requirements of the species involved, reserve design, and monitoring, and it should also include advice on descriptions of management options and application of risk analyses in consideration of alternatives.

---

[6]Estimates of periods in years are important for planning of human activities, and generation times are important to provide biological realism to the estimates.

REFERENCES

Andrewartha, H.G., and L.C. Birch. 1954. The Distribution and Abundance of Animals. Chicago: University of Chicago Press.

Beatley, T. 1994. Habitat Conservation Planning. Austin, Tex.: University of Texas Press.

Clements, F.E., and V.E. Shelford. 1939. Bio-Ecology. New York: John Wiley & Sons.

DOI (U.S. Department of the Interior) Office of the Inspector General. 1990. Audit Report: The Endangered Species Program. U.S. Department of the Interior, Washington, D.C.

Elton, C. 1927. Animal Ecology. London: Sedgwick and Jackson.

FWS (U.S. Fish and Wildlife Service). 1990. Interim National Conservation Planning Guidelines. U.S. Department of the Interior, U.S. Fish and Wildlife Service, Washington, D.C.

FWS (U.S. Fish and Wildlife Service). 1992. Report to Congress: Endangered and Threatened Species Recovery Program. U.S. Department of the Interior, U.S. Fish and Wildlife Service, Washington, D.C.

FWS (U.S. Fish and Wildlife Service). 1994. Preliminary Draft Handbook for Habitat Conservation Planning and Incidental Take Permit Processing. U.S. Department of the Interior, U.S. Fish and Wildlife Service, Washington, D.C.

Gilbert, L.E. 1980. Food web organization and the conservation of neotropical diversity. Pages 11-33 in M.E. Soulé and B.A. Wilcox, eds. Conservation Biology: An Evolutionary-Ecological Perspective. Sunderland, Mass.: Sinauer Associates.

Groombridge, B., ed. 1992. Global Biodiversity: Status of the Earth's Living Resources. New York: Chapman and Hall.

Jackson, T.C. 1992. All creatures great and small. Legal Times, Dec. 7, pp. 20-23.

Jensen, D.B., M.S. Torn, and J. Harte. 1993. In Our Own Hands: A Strategy for Conserving California's Biodiversity. Berkeley, Calif.: University of California Press.

Jones, G.E. 1987. The Conservation of Ecosystems and Species. New York: Croom Helm.

Krebs, C.J. 1985. Ecology: Experimental Analysis of Distribution and Abundance. 3rd Ed. New York: Harper & Row.

Mann, C.C., and M.L. Plummer. 1995. Noah's Choice: The Future of Endangered Species. New York: Alfred A. Knopf.

McArthur, R.H., and E.O. Wilson. 1967. The Theory of Island Biogeography. Princeton, N.J.: Princeton University Press.

McNaughton, S.J. 1989. Ecosystems and conservation in the twenty-first century. Pp 109-120 in Conservation for the Twenty-First Century, D. Western and M.C. Pearl, eds. New York: Oxford University Press.

NRC (National Research Council). 1993. Setting Priorities for Land Conservation. Washington, D.C.: National Academy Press.

Pianka, E. 1978. Evolutionary Ecology, 2nd Ed. New York: Harper & Row.

Tear, T.H., J.M Scott, P.H. Hayward, and B. Griffith. 1993. Status and prospects for success of the Endangered Species Act: A look at recovery plans. Science 262:976-977.

Thornton, R.D. 1991. Searching for consensus and predictability: Habitat conservation planning under the Endangered Species Act of 1973. Environ. Law 21:605-656.

Van Horne, B. 1983. Density as a misleading indicator of habitat quality. J. Wildl. Mgmt. 47:893-901.

Whitford, P.B. 1949. Distribution of woodland plants in relation to succession and clonal growth. Ecology 30:199-208.

Wilson, E.O. 1992. The Diversity of Life. Cambridge, Mass.: Harvard University Press.

5

Modern Perspectives of Habitat

Wild species cannot exist in isolation from their habitats. Although scientists and lay persons often speak of species as though they were independent entities, organisms and habitats are inextricably linked. In his classic textbook, Elton (1927) recognized that the identification and naming of a species in essence coded the species for its habitat and ecological role. He encapsulated this insight in the term "niche." In a real sense, then, species and habitats are but elements of a single ecological system. This insight is echoed in the concept of the ecosystem, which was introduced to describe the intimate linkage of organisms with their habitats and the physical conditions, resources, and other interacting organisms in those habitats. Thus, an ecosystem was viewed as the assemblage of organisms and the physical environment with which they exchange energy and matter. That the system is fully linked is a subtle and powerful concept that sometimes even ecologists fail to recognize or appreciate. The ecosystem concept demands that we understand the reciprocal (not necessarily always beneficial or mutualistic) relationships between organisms and their habitats. The Endangered Species Act encapsulates part of this wisdom in its explicit statement of purpose "to conserve the ecosystems upon which threatened and endangered species depend."

Recognizing that species and habitats are two components of a single system puts the language of the ESA in a specific scientific light. Although a substantial portion of the language of the ESA focuses on the organismal part of the species-habitat system (harassing, prohibitions against the killing or pursuit of listed species, etc.), the term "harm" biologically should encompass damage to the entire system, including the physical components of the system, through damage to any of its parts. Such is the nature of systems organization (e.g., von Bertalanffy, 1968; Pattee, 1973; Kolasa and Pickett, 1989). Therefore, harm in an ecological sense applies to damage to the habitat of a species or curtailment of a species' access to a habitat.

Species and habitat conservation would seem to be about saving living things. However, the issue is not so straightforward. Populations with identifiably distinct evolutionary and ecological features that exist at any one time and across a certain space are not static entities. They are part of ecological and evolutionary streams that stretch into the past and have at least some biological potential for continuing into the future. The evolutionary stream for one species can interact with other evolutionary streams by sharing genetic material or by dividing to produce separate lineages. Viewed from this dynamic perspective, conservation of species requires the conservation of ongoing evolutionary processes and potential. Species, as diagnosed at any particular time, represent a sampling of a continuous evolutionary process.

As members of communities and ecosystems, species take part in ecological processes. These ecological processes can also be thought of as streams, including the dynamics of community succession, the rhythm of natural disturbance, the waxing and waning of predator and prey populations, and the cycling of soil nutrients. To protect species, their ecological streams must also be protected.

Habitats that support these evolutionary and ecological streams are heterogeneous, and that heterogeneity is expressed in space and in time. Like the profound significance of habitat, this too is one of the fundamental concepts of modern ecology (Wiens, 1976; Kolasa and Pickett, 1991). Habitats (or habitat diversity) should be viewed both spatially and temporally if conservation planning is to be successful. Spatial heterogeneity takes several common forms, ranging from gradual to discrete (Kolasa and Rollo, 1991). Gradients of heterogeneity are patterns established by gradual change in such factors as moisture, nutrients and prey, temperature, topography, soil chemistry, exposure to the elements, and cover from predators (Whittaker, 1975; Austin, 1985; Schoener, 1986). Ecotones are coarse-scale gradients representing the transitional boundaries of biomes or community types, and are governed by spatial changes in climatic and human land-use practices (Holland et al., 1991). On finer scales, spatial gradients can be generated at the edges of communities having contrasting structures or compositions (Gosz, 1991).

In contrast with the pattern illustrated by gradients, habitat heterogeneity can be patchy in its configuration. Patches are discrete spatial units that are detectable on a certain scale. Patches may result from precipitous changes in physical environment or biological composition through space (Forman and Godron, 1986; Forman, 1987). For example, a pond is a patch recognizable on a topographical map. On the same map, other patches, such as vegetated land, wetlands, human settlements, or hedgerows in an agricultural matrix, might be discernible. Patches can arise from fixed spatial patterns of resources, soil features, or topographic features. More mutable patches may reflect human land uses, biological interactions, natural or human disturbances, and successions. The fact that biological interactions and successions can form patches means that patches in an area can differ in age as well as origin and can change over time.

Patchiness can reflect the underlying physical characteristics of a site and the biological environment that emerges from that physical environment and from the interactions of species. Thus, the distribution, productivity, and assemblages of organisms are themselves sources of heterogeneity above and beyond that which is caused directly by the physical environment (Schoener, 1986). Organisms in one type of patch might depend in turn on organisms or processes that depend on contrasting patch types (e.g., pollinators of certain tropical vines) (Gilbert, 1980). Each organism responds to the physical and biotic heterogeneity it is exposed to in an individualistic fashion, based on its unique combination of resource requirements, physical tolerances, and capacities for biotic interaction (Austin, 1985; Shipley and Keddy, 1987).

The spatial pattern in the physical and biological environment and species' responses to them are not static. Rather, change in the environment and, consequently, change in the distribution, growth, abundance, and interaction of species are common. Habitat can shift in distribution or be unoccupied for a time (Levins, 1970; Horn and MacArthur, 1972). A landscape perspective highlights the fluxes between patches and provides critical scientific information about the contexts on which species depend (Pickett and Thompson, 1978; Noss, 1987a, 1987b; Angelstam, 1992; Fiedler et al., 1993). Organisms might encounter habitat as shifting mosaics or dynamic arrays of patches for two reasons. First, habitat can shift in location through time due to a variety of changes in environment; natural changes in climate are a major source of habitat shifts (Neilson and Wullstein, 1983). In addition, habitat can be created or destroyed by episodic or rare events, such as fire or windstorms (Garwood et al., 1979; White, 1979). Habitat destruction means that a site is converted from an environment suitable to a species to an environment that is adverse to that species. Furthermore, habitat destruction for one species might constitute habitat enhancement for another. Episodic natural events that can alter habitat include physical disturbances, such as fire, windthrow, mass movements, flooding, and outbreaks of diseases or herbivorous insects (Pickett and White, 1985).

Second, habitat can be unoccupied for a time due to migrations or movements of organisms in response to seasonal cycles, reproductive behavior, localized resource depletion or creation, and a search for protection (Rotenberry and Wiens, 1980; Angelstam, 1992). Areas through which a species might travel for any of the reasons above should be considered a part of its habitat as much as areas continuously occupied by a species (Noss, 1991).

A landscape perspective can provide an encompassing view of the dynamics of habitat. A landscape is a large area in which a certain array of ecosystem types is linked by natural disturbance regime, pattern of human land use and disturbance, and distribution of land forms (Forman and Godron, 1986; Risser, 1987). Recovery planning for threatened and endangered species that reflects a landscape view of habitat is more likely to be successful than planning that ignores it (Franklin, 1993).

LANDSCAPES AND POPULATIONS

Populations often occur as collections of relatively discrete demographic units distributed over a landscape. Such subdivided populations, in which individual demographic units are connected through dispersal or migration, are referred to as metapopulations (Levins, 1970). The concept has had its widest use in evolutionary biology, where the movement of genes and the potentially differential effects of natural selection among the populations are important mechanisms for evolutionary change or stasis (Soulé and Wilcox, 1980). Whether demographic units are genetically differentiated and the degree of gene flow that occurs among the populations are important concerns in evolutionary metapopulation dynamics.

Metapopulation concepts are applicable to ecological concerns as well as to evolution. In an ecological context, concerns are with 1) the degree of landscape heterogeneity, 2) the degree to which population structure reflects environmental heterogeneity, and 3) the demographic, community, and ecosystem consequences of population subdivision (Hanski, 1982; Wiens, 1984). In addition, environmental heterogeneity can change through time, and metapopulation dynamics might respond to those changes. The local dynamics of a population can be determined in large part by influences from adjacent patches in a landscape (Angelstam, 1992). As a consequence of all the interactions mentioned above, metapopulation structure and dynamics might be best examined through an inclusive and dynamic view of habitat (Pulliam, 1988). Specific sites become more or less suitable as habitat, reflecting a wide variety of factors. Furthermore, the environmental heterogeneity that affects the persistence and performance of a species can change through time (Pickett, 1976; Pickett and Thompson, 1978). A concept that recognizes environmental heterogeneity—that of "source" and "sink" habitats—should be incorporated into conservation planning.

SOURCES AND SINKS

In natural populations, individuals reside in habitat patches of differing quality. Individuals in highly productive habitats can be expected to be more successful in producing offspring than those in poor habitats, which can be expected to suffer poor reproductive success or survival. The fate of a population as a whole can depend on whether the reproductive success of the individuals in high-value habitats outweighs the failure of the individuals in the poor areas. This concept has its own nomenclature. Sources are areas where local reproductive success is greater than local mortality. Populations in source habitats produce an excess of individuals, which disperse outside their natal habitat patch to find a place to settle and to breed. In contrast, sinks are habitat areas where local productivity is less than local mortality; in the absence of immigration from source areas, populations in sink habitats decline toward extinction.

Sources and sinks can be defined in reference to the finite rate of increase (λ) for the population in a given area. The finite rate of increase can change across either space or time as the survival rates or reproductive rates of the population vary. The geometric mean $\underline{\lambda}$ of the rates for a sequence of t years characterizes the mean growth rate. When the long-term mean growth rate ($\underline{\lambda}$) is less than 1.0, the population will decline, but if $\underline{\lambda}$ exceeds 1.0, the population will grow.

The finite rate of increase can also be used to describe spatial variation in population growth rates by calculating λ based on the birth and death rates that apply in a specific habitat or patch of habitat (Pulliam, 1988). If patches of habitat are isolated from one another, the finite rate of increase for each habitat patch describes the growth rate in that patch; however, when habitat patches are less isolated, the concept of habitat-specific growth rate is complicated by dispersal. When habitats are connected by dispersal, different parts of the population are growing at different rates. The growth rate and the habitat-specific λ_i gives the rate that the population in habitat patch i should grow in the absence of immigration or emigration. The finite rate of increase of the entire interconnected population is given by the weighted average of the λs across all habitats (λ_i is weighted by n_i/N, where n_i is the population size in the ith habitat, and N is the total population size over habitats.)

Large numbers of individuals live in sink habitats, despite the potential disappearance of the sink population without immigration from more productive areas. To illustrate this point, assume that the subpopulation in the source grows at the rate λ_1 until it reaches a maximum size (n_1^*), which represents the maximum number of breeding individuals that can be accommodated in the source. Once the source has reached its maximum size ($\lambda_1 n_1^*$ individuals at the end of each season), only n_1^* can remain to breed;

the remaining $n_1^*(\lambda_1 - 1)$ are assumed to emigrate from the source habitat into sink habitat. In the absence of immigration, the sink subpopulation would soon disappear. However, with a steady immigration of individuals from the source habitat, the sink population will grow to an equilibrium population of $n_2^* = n_1^*(\lambda_1 - 1)/(1 - \lambda_2)$. Clearly, if the per capita reproductive surplus $(\lambda_1 - 1)$ in the source is much larger than the reproductive deficit $(1 - \lambda_2)$ in the sink, then the sink habitat will contain substantially more individuals than the source habitat, despite the fact that the sink subpopulation is dependent on emigration from the source for its very existence. This example illustrates a more general conclusion that the majority of individuals in a local population likely exist in habitat that is unsuitable for them over the long term (Pulliam, 1988).

It is important to remember that source habitat is defined by demographic characteristics—habitat-specific reproductive success and survivorship—and not by population density; therefore, population density can be a misleading indicator of habitat quality (van Horne, 1982). Source habitats could easily be overlooked if conservation efforts concentrate only on habitats where a species is most common, rather than where it is most productive. If source habitats are not protected by conservation plans, an entire metapopulation could be threatened.

Environmental heterogeneity on the landscape scale can be represented as a mosaic. Changes in the kinds or arrangement of patches in a landscape result in a shifting mosaic (Botkin and Sobel, 1975; Bormann and Likens, 1979). Underlying such mosaics are the various mechanisms of patch dynamics, which include natural disturbance, life-history features of organisms, and succession (Pickett and White, 1985; Walker, 1989; Luken, 1990), as well as anthropogenic changes. Source-sink relations likely exist in such mosaic landscapes. A step toward incoporating the landscape perspective of habitat is to broaden the analysis to the concept of the metapopulation.

METAPOPULATIONS

Metapopulation is a more encompassing concept than that of source and sink dynamics, because demographic rates in metapopulations might not be the same in different patches of habitat. Source and sink dynamics are a special case of metapopulation dynamics in which some habitat patches (sources) are substantially better than others (sinks).

Levins (1969) argued that the fraction of suitable habitat patches that are occupied at any time represents a balance of the rates at which subpopulations go extinct in occupied patches and the rates of colonization of empty patches (see Hanski, 1989). The rate of local extinction depends on conditions within a patch and the stochastic nature of the dynamics of small populations. The rate of colonization of empty patches depends on the dispersal ability of the species and the location of suitable patches in the landscape (Hansson et al., 1992).

Metapopulation models can be used to describe the structure and dynamics of populations that are scattered across a landscape in spatially isolated patches. Such models are useful in the identification of particular subpopulations, habitat patches, or links between patches that are critical to the maintenance of the overall metapopulation. Beier's (1993) study of cougars (*Felis concolor*) in the Santa Ana Mountains of southern California provides an excellent example of this type of analysis. Beier used radiotelemetry data to show that the California cougars exist in a collection of semi-isolated populations found mostly in small mountain ranges linked by riparian corridors. He developed a metapopulation simulation model and showed that the metapopulation of the region was heavily dependent on movement by individual cougars through the corridors to colonize empty areas. Beier's analysis quantified how the loss of habitat in this region and corresponding decrease in population size would affect the chance of extinction for the entire metapopulation. For example, by examining the importance of specific patches and corridors in maintaining the metapopulation, Beier showed that one corridor in the northern part of the study area linked a 150-km^2 patch (8% of the total area) with the rest of the region. Recognition of the roles of habitat heterogeneity, habitat patchiness, the importance of suitable but temporarily unoccupied habitat, and the distinctions between source and sink habitats are all critical to successful implementation of the ESA. Recent advances in metapopulation theory and the modeling of demographic phenomena offer managers the tools necessary to plan better for conservation of species and the habitats upon which they depend. Spatially explicit models incorporate these critical concepts.

SPATIALLY EXPLICIT MODELS

Landscape ecology and conservation biology have made clear that the geometry of habitat patches in a landscape can influence population trends and extinction probabilities. Metapopulation models have generally ignored the complexities of dispersal behavior and habitat geometry by assuming that individuals are equally likely to disperse to near and distant sites. The models are useful for gaining general insights into population dynamics but not for managing particular species on real landscapes. Such models should incorporate landscape patterns that determine spatial patterns in populations.

Disturbance opens new patches in a landscape. Fires or windstorms open communities, alter resources, and kill existing organisms; newly arrived organisms or seeds and spores respond to the disturbed sites (Grubb, 1977). Life-history phenomena, such as rates of growth, maximum longevity, change in growth form with age, and onset and temporal patterns of reproduction, can all affect the origin and disappearance of patches in landscapes (Thompson, 1982). Likewise, the interaction of organisms that leads to successional change in community composition or structure through time alters the distribution of patches in a landscape (Foster, 1980). Because various areas in a landscape can undergo disturbance, planning must accommodate infrequent events (Pickett and Thompson, 1978; NRC, 1993). Disturbances are either biotic, as with diseases, or abiotic, as with windstorms. If disturbance is likely to obliterate or reduce significantly the density of a rare species in certain patches in a landscape, other patches must remain occupied to permit recolonization of the disturbed patches (Shafer, 1990).

Species distributions can vary dramatically through time owing to patch dynamics and shifting mosaics, and so species might require more sites over the long term than is apparent from their distribution at any one time. A long-term perspective is required to understand habitat requirements thoroughly. Dispersal patterns and mechanisms become critical aspects of species biology in situations where movement among patches is mandated by patch dynamics (McNaughton, 1989; Hansson et al., 1992). Dispersal can be episodic or continuous and can require available sites for species to move through—sites required for dispersal are part of the habitat of a species. Such dispersal habitat can be arrayed as stepping stones or unbroken corridors (McDonnell and Pickett, 1988). Stepping-stone patterns are exemplified by the dispersal habitat used by migrating waterfowl or other birds, and the more readily recognized corridor patterns are exemplified by small mammals (Merriam and Lanoue, 1990).

Patch dynamics involve another kind of habitat as well. In addition to habitat that organisms occasionally or periodically disperse through, refuge areas might be needed. During periods of physical environmental stress or unusually intense or large disturbances, organisms might be extirpated from their usual or customary habitat (Pickett and Thompson, 1978). Unless organisms have areas in which they can temporarily find shelter or in which seeds, larvae, or adults can persist through disturbances and stresses, the long-term persistence of a species will be compromised. Refuges and recolonization sources thus become an important aspect of habitat (Ås et al., 1992). The possibility that organisms would require dispersal patches or corridors, refuges, or recolonization sources demands that a habitat plan explicitly include such areas. Likewise, organisms might depend on others (e.g., for food or dispersal) that require additional habitats than those in which the interactions take place (Gilbert, 1980).

Spatially explicit population models are well suited for encompassing realistic details of particular species and landscapes into conservation plans. Spatially explicit models incorporate the actual location of suitable patches of habitat and explicitly consider the movement of organisms among such patches. For example, Mobile Animal Population (MAP) is a class of spatially explicit population-simulation models (Liu, 1992; Pulliam et al., 1992) that incorporates changes in land-use and habitat availability, habitat-specific demography, and the dispersal behavior of organisms in computer representations of real landscapes. In MAP models, landscapes are represented as grids of cells and clusters of adjacent cells that represent the size and location of habitat patches in mosaic landscapes. MAP models contain subroutines that specify, for example, forest-management practices, succession, and other aspects of forest dynamics. MAP models can depict the current landscape structure and project that landscape structure into the future based on a management plan specifying a harvest and replanting schedule. Management activities, such as thinning or controlled burning of stands, which might influence stand suitability for particular species of interest, can be easily incorporated into MAP models. MAP models can run on landscape maps generated by geographic information systems, which incorporate the actual distribution of habitat patches in a region.

Although spatially explicit models are a new development, they are beginning to be used as land-management and planning tools. Spatially explicit population models developed for the spotted owl have proven useful in the Pacific Northwest and California (Verner et al., 1992; McKelvey et al., 1993). An analysis using the spatially explicit model for the spotted owl has identified specific owl populations in the San Gabriel and San Bernardino Mountains as being critical to the viability of an entire southern California metapopulation (Verner et al., 1992). In another application of spatially explicit landscape models, Turner et al. (1994) developed a spatially explicit model for wintering herds of bison and elk in Yellowstone National Park. That model has been used to explain how bison and elk have responded to the local patterns of habitat diversity caused by the large-scale Yellowstone fires of 1988 and should be useful in future land-use management and fire-control planning.

One of the best studies of patch dynamics for a plant species was carried out by Menges and coworkers on Furbish's lousewort (*Pedicularis furbishiae*) before the recent development of spatially explicit models (Menges et al., 1986; Menges, 1990). This plant is an herbaceous perennial species endemic to the Saint John River Valley in Northern Maine. Furbish's lousewort exists in very unstable habitat patches along the banks of the Saint John River. Menges describes this lousewort as inhabiting "a disturbance/successional niche" defined by hydrology and vegetation response. The species is a poor competitor and seems to do best in low riverbank sites characterized by nonwoody vegetation, frequent flooding, and springtime ice scour. Menges and his colleagues measured habitat-specific demographic variables and concluded that in the absence of catastrophic disturbance, only wet, early successional sites can maintain viable populations. The system is characterized by catastrophic events that lead to local extinction, such as ice scour and bank slumping. Menges concluded that local population extinction probability was high even in the best of sites, stating that "individual *P. furbishiae* populations are temporary features of the riverine ecosystem" and that metapopulation viability depends on a positive balance between new populations and extinction (Menges, 1990).

The above examples of models and studies are sufficiently well developed or defined to be used in conservation and management. But for most species, the relevant details of population biology necessary for conservation planning are not known, and years of concentrated field work would be required to parameterize the models. However, a wide array of new ecological concepts and information can be applied to the conservation and recovery of endangered species. Since the passage of the ESA in 1973, a variety of new ecological tools have been developed that can help plan and manage subdivided populations in spatially heterogeneous and dynamic landscapes.

A SPATIAL PERSPECTIVE AND POPULATION VIABILITY ANALYSIS

Planning for habitat and population management must account for metapopulation structure, both from a genetic and an ecological perspective. Several specific mechanisms are required to maintain dispersed populations in a landscape. Spatial connections between populations must be permitted to continue. Depending on the nature of the connections between different subpopulations or landscape patches, contiguous habitat or dispersal must be allowed or encouraged (Noss, 1983; Noss and Harris, 1986). Habitat-connecting corridors for subpopulations in a landscape might not be continuously occupied.

Determining which habitats are sources and which are sinks requires detailed field studies and a great deal of knowledge about the natural history of the organisms of concern. Simple measures of densities can be inadequate to expose source-sink dynamics. A rigorous analysis of source-sink dynamics requires information on birth and death rates of individuals in each habitat type and some knowledge of dispersal behavior of the organism. And although studies to obtain such basic information are critical for managing population viability, preliminary conservation strategies can be formulated without detailed estimates of needed details of organisms' biology, and models can be updated as more information becomes available.

Consideration of source-sink dynamics is an important aspect of reserve design and habitat protection. In some cases, adding additional habitat to a reserve actually results in a smaller metapopulation, if most of the additional land is sink habitat (Pulliam and Danielson, 1991). Individuals dispersing within a reserve might settle in the unproductive sink patches if the available source patches are too hard to find. Recent studies using the metapopulation model developed for spotted owls predict such a problem with some reserve designs proposed for the species in the Pacific Northwest (McKelvey et al., 1993).

Population Viability Analysis (PVA) provides an inclusive technique that can accommodate many of the insights from the modern ecological view of habitat as a landscape phenomenon. PVA is concerned with how habitat loss, environmental uncertainty, demographic stochasticity, and genetic factors interact to determine extinction probabilities for individual species (Soulé, 1987; Shafer, 1990). Though PVA is a relatively new approach, several excellent studies have demonstrated its usefulness (e.g., Ehrlich and Murphy, 1987; Marcot and Holthausen, 1987; Menges, 1990; Murphy et al., 1990; Stacey and Taper, 1992).

Many ecological factors that influence the likelihood of population extinction can be incorporated into PVA. These include (1) demographic stochasticity, (2) environmental uncertainty, (3) natural catastrophes, and (4) genetic uncertainty.

As a rough rule of thumb, genetic and demographic uncertainty are important factors only in small populations, or populations that have low effective population sizes despite relatively high actual census sizes (see Chapter 7). Environmental uncertainty and catastrophes can affect the viability of much larger populations. Conservation strategies and recovery planning often must deal with the combined effects of all four factors, because many endangered species, especially large vertebrates, exist in small populations. The recovery plans for endangered species should usually employ two goals for promoting viable populations: creation of multiple populations, so that a single catastrophic event cannot destroy the whole species, and increasing the size of each population to a level where the threats of genetic, demographic, and normal environmental uncertainties are diminished. Any attempt to determine population viability must be done with an understanding that predictions are made in a context of uncertainty (see Chapter 7).

Most PVAs to date have combined field studies on important demographic parameters and simulation modeling on the possible effects of various extinction factors. Generally, the object of the analyses is to generate a prediction of the probability that a population will become extinct in a given number of years (e.g., a 95% probability of extinction within 100 years). Murphy et al. (1990) suggested that species fall along a continuum between two extremes:

• Organisms, such as most large vertebrates, with low population densities that are comparatively widespread (most endangered large vertebrates, for example). PVAs for such species should focus on the genetic and demographic factors that affect especially small populations. (This is the style of PVA that has been done most frequently.)

• Organisms, such as most invertebrates and small vertebrates, that are frequently restricted to few habitat patches, but within those patches can reach high population densities. PVAs for those species must emphasize environmental uncertainty and catastrophic factors.

Extinction due to environmental and catastrophic stochasticity is more important in small populations, so all factors need to be taken into account in such situations. This is not to say that some factors will not be more important than others in specific cases.

Many human-induced and other changes in landscape and ecosystem function can be slow to become apparent, especially when long-lived species are involved. For example, the long-lived razorback sucker (*Xyrauchen texanus*) remained common in impoundments of the lower Colorado River for many years although no juvenile suckers were found there (Ono et al., 1983). As another example, fire suppression can take a long time to produce effects in an ecosystem. This means that conclusions from PVAs and management based on them should be viewed with caution.

CONCLUSIONS

• Assessing a conservation and habitat plan must take a retrospective view in many situations. In some cases, metapopulation dynamics in human-populated landscapes suffers from the absence of processes that previously contributed to maintaining the species population (see Ehrlich and Murphy, 1987). Such processes might include succession, disturbance, predation, mutualism, and the like. Processes that have been lost must be replaced, substituted, or compensated for if the species is to be maintained (Walker, 1989; Wagner and Kay, 1993).

• Management and planning for metapopulation dynamics in landscapes must be monitored to determine their effectiveness, because the conditions in the landscape might change, or the management may not be as effective as initially thought (Barrett, 1985; National Research Council, 1986; Schroeder and Keller, 1990; Irwin and Wigley, 1993). The status of the component populations must be assessed at intervals. The monitoring interval will be determined by the longevity and generation time of the organism of interest or the expected periodicity of rare events and episodic interactions in which the species is involved. Monitoring must also assess the condition of the occupied habitat and the habitat necessary for dispersal (Hansson, 1992).

• Monitoring will indicate the effectiveness of a management strategy. If the management does not maintain an occupied or dispersal habitat in suitable condition for a species, then the tactics and environmental components targeted by the management can be adjusted (Schroeder and Keller, 1990). This strategy of monitoring the results of management to assess the appropriateness and success of the strategy and to adjust it if necessary is labeled *adaptive management*, a particularly appropriate term, considering the environments of species can undergo many natural and anthropogenic changes. Such changes can be rapid and unexpected or gradual and difficult to detect. In either event, the changes can have untoward results for a target species, necessitating adjustments in conservation efforts.

• A second characteristic of successful planning for maintenance of species is including information on the interactions in which they engage. All species exist as parts of food webs and interaction networks (McNaughton, 1989; Pimm, 1991). Interactions include those with prey and resources, potential mates, consumers, competitors, pollinators and dispersers. Management without attention to networks of interaction will fail to maintain critical resources or constraining factors in the species' environment (Holt and Talbot, 1978). Management that accommodates the interaction networks is labeled *ecosystem management* (see Chapters 9 and 10). Ecosystem management involves a turn from the focus on management for commodities only (Jones, 1987; Hartshorn and Pariona-A, 1993) and focuses instead on the ecosystem processes of population, community, and biogeochemical interactions to maintain the condition and function of a site as a whole (Likens, 1992; Society of American Foresters, 1993).

REFERENCES

Angelstam, P. 1992. Conservation of communities—The importance of edges, surroundings and landscape mosaic structure. Pp. 9-70 in Ecological Principles of Nature Conservation: Applications in Temperate and Boreal Environments, L. Hansson, ed. New York: Elsevier Applied Science.

Ås, S., J. Bengtsson, and T. Eberhard. 1992. Archipelagoes and theories of insularity. Pp. 201-251 in Ecological Principles of Nature Conservation: Applications in Temperate and Boreal Environments, L. Hansson, ed. New York: Elsevier Applied Science.

Austin, M.P. 1985. Continuum concept, ordination methods, and niche theory. Annu. Rev. Ecol. Syst. 16:39-61.

Barrett, G.W. 1985. A problem-solving approach to resource management. BioScience 35:423-427.

Beier, P. 1993. Determining minimum habitat areas and habitat corridors for cougars. Conserv. Biol. 7:94-108.

Bormann, F.H., and G.E. Likens. 1979. Catastrophic disturbance and the steady-state in northern hardwood forests. Am. Sci. 67:660-669.

Botkin, D.B., and M.J. Sobel. 1975. Stability in time-varying ecosystems. Am. Naturalist 109:625-646.

Ehrlich, P.R., and D.D. Murphy. 1987. Conservation lessons from long-term studies of checkerspot butterflies. Conserv. Biol. 1:122-131.

Elton, C. 1927. Animal Ecology. London: Sedgwick and Jackson.

Fiedler, P.L., R.A. Leidy, R.D. Laven, N. Gershenz, and L. Saul. 1993. The contemporary paradigm in ecology and its implications for endangered species conservation. Endangered Species Update 10:7-12.

Forman, R.T.T. 1987. The ethics of isolation, the spread of disturbance, and landscape heterogeneity. Pp. 213-229 in Landscape Heterogeneity and Disturbance, M.G. Turner, ed. New York: Springer.

Forman, R.T.T., and M. Godron. 1986. Landscape Ecology. New York: John Wiley & Sons.

Foster, R.B. 1980. Heterogeneity and disturbance in tropical vegetation. Pp. 75-92 in Conservation Biology: An Evolutionary-Ecological Perspective, M.E. Soulé and B.A. Wilcox, eds. Sunderland, Mass.: Sinauer Associates.

Franklin, J.F. 1993. Preserving biodiversity: species, ecosystems, or landscapes. Ecol. Appl. 3:202-205.

Garwood, N.C., D.P. Janos, and N. Brokaw. 1979. Earthquake caused landslides: A major disturbance to tropical forest. Science 205:997-999.

Gilbert, L.E. 1980. Food web organization and the conservation of neotropical diversity. Pp. 11-33 in Conservation Biology: An Evolutionary-Ecological Perspective, M.E. Soulé and B.A. Wilcox, eds. Sunderland, Mass.: Sinauer Associates.

Gosz, J.R. 1991. Fundamental ecological characteristics of landscape boundaries. Pp. 8-30 in Ecotones: The role of Changing Landscape Boundaries in the Management and Restoration of Changing Environments. New York: Chapman and Hall.

Grubb, P.J. 1977. The maintenance of species-richness in plant communities: The importance of the regeneration niche. Biol. Rev. 52:107-145.

Hanski, I. 1982. Dynamics of regional distribution: the core and satellite species hypothesis. Oikos 38:210-221.

Hanski, I. 1989. Metapopulation dynamics: Does it help to have more of the same? Trends Ecol. Evol. 4:113-114.

Hansson, L. 1992. Landscape ecology of boreal forests. Trends Ecol. Evol. 7:229-302.

Hansson, L., L. Söderström, and C. Solbreck. 1992. The ecology of dispersal in relation to conservation. Pp. 162-200 in Ecological Principles of Nature Conservation: Applications in Temperate and Boreal Environments, L. Hansson, ed. New York: Elsevier Applied Science.

Hartshorn, G.S., and W. Pariona-A. 1993. Ecologically sustainable forest management in the Peruvian Amazon. Pp. 151-166 in Perspectives on Biodiversity: Case Studies of Genetic Resource Conservation and Development, C.S. Potter, J.I. Cohen, and D. Janczewski, eds. American Association for the Advancement of Science, Washington, D.C.

Holland, M.M., P.G. Risser, and R.J. Naiman, eds. 1991. Ecotones: The Role of Landscape Boundaries in the Management and Restoration of Changing Environments. New York: Chapman and Hall.

Holt, S.J., and L.M. Talbot. 1978. New Principles for the Conservation of Wild Living Resources, Vo. 59. Wildlife Society, Louisville, Ky.

Horn, H.S., and R.H. MacArthur. 1972. Competition among fugitive species in a harlequin environment. Ecology 53:749-752.

Irwin, L.L., and T.B. Wigley. 1993. Toward an experimental basis for protecting forest wildlife. Ecol. Appl. 3:213-217.

Jones, G.E. 1987. The Conservation of Ecosystems and Species. New York: Croom Helm.

Kolasa, J., and S.T.A. Pickett. 1989. Ecological systems and the concept of organization. Proc. Natl. Acad. Sci. USA 86:8837-8841.

Kolasa, J., and S.T.A. Pickett, eds. 1991. Ecological Heterogeneity. New York: Springer.

Kolasa, J., and C.D. Rollo. 1991. Introduction: The heterogeneity of heterogeneity: A glossary. Pp. 1-23 in Ecological Heterogeneity, J. Kolasa and S.T.A. Pickett, eds. New York: Springer.

Levins, R. 1969. Some demographic and genetic consequences of environmental heterogeneity for environmental control. Bull. Entomol. Soc. Am. 15:237-240.

Levins, R. 1970. Extinction. Pp. 77-107 in Some Mathematical Questions in Biology, Vol. 2, M. Gerstenhaber, ed. American Mathematical Society, Providence, R.I.

Likens, G.E. 1992. Excellence in Ecology. 3: The Ecosystem Approach: Its Use and Abuse. Ecology Institute, Oldendorf/Luhe, Germany.

Liu, J. 1992. ECOLECON: A Spatially Explicit Model for Ecological Economics of Species Conservation in Complex Forest Landscapes. Ph.D. Dissertation. University of Georgia, Athens, Ga.

Luken, J.O. 1990. Directing Ecological Succession. New York: Chapman and Hall.

Marcot, B.G., and R. Holthausen. 1987. Analyzing population viability of the spotted owl in the Pacific Northwest. Trans. North Am. Wildl. Nat. Res. Conf. 52:333-347.

McDonnell, M.J., and S.T.A. Pickett. 1988. Connectivity and the theory of landscape ecology. Münstersche Geographische Arbeiten 29:17-21.

McKelvey, K., B.R. Noon, and R.H. Lamberson. 1993. Conservation planning for species occupying fragmented landscapes: The case of the northern spotted owl. Pp. 424-450 in Biotic Interactions and Global Change, P.M. Kareiva, J.G. Kingsolver, and R.B. Huey, eds. Sunderland, Mass.: Sinauer Associates.

McNaughton, S.J. 1989. Ecosystems and conservation in the twenty-first century. Pp. 109-120 in Conservation for the Twenty-First Century, D. Western and M.C. Pearl, eds. New York: Oxford University Press.

Menges, E. 1990. Population viability analysis for an endangered plant. Conserv. Biol. 4:52-62.

Menges, E., D.M. Waller, and S.C. Gawler. 1986. Seed set and seed predation in *Radicularis furbishiae*, a rare endemic of the St. Johns River, Maine. Am. J. Bot. 73:1168-1177.

Merriam, G., and A. Lanoue. 1990. Corridor use by small mammals: Field measurement for three experimental types of *Peromyscus leucopus*. Landscape Ecol. 4:123-131.

Murphy, D.D., K.E. Freas, and S.B. Weiss. 1990. An environment-metapopulation approach to population viability analysis for a threatened invertebrate. Conservation Biology 4(1):41-52.

NRC (National Research Council). 1986. Ecological Knowledge and Environmental Problem-Solving: Concepts and Case Studies. Washington, D.C.: National Academy Press.

NRC (National Research Council). 1993. Setting Priorities for Land Conservation. National Research Council, Washington, D.C.

Neilson, R.P., and L.H. Wullstein. 1983. Biogeography of two southwest American oaks in relation to atmospheric dynamics. J. Biogeogr. 10:275-297.

Noss, R.F. 1983. A regional landscape approach to maintain diversity. BioScience 33:700-706.

Noss, R.F. 1987a. From plant communities to landscapes in conservation inventories: A look at The Nature Conservancy (USA). Biol. Conserv. 41:11-37.

Noss, R.F. 1987b. Protecting natural areas in fragmented landscapes. Natural Areas J. 7:2-13.

Noss, R.F. 1991. Landscape connectivity: Different functions at different scales. Pp. 27-39 in Landscape Linkages and Biodiversity, W. E. Hudson, ed. Washington, D.C.: Island Press.

Noss, R.F., and L.D. Harris. 1986. Nodes, networks, and MUMs: Preserving diversity at all scales. Environ. Manage. 10:299-309.

Ono, R.D., J.D. Williams, and A. Wagner. 1983. Vanishing Fishes of North America. Washington, D.C.: Stone Wall.

Pattee, H.H. 1973. The physical basis and origin of hierarchical control. Pp. 71-108 in Hierarchy Theory: The Challenge of Complexity, H.H. Pattee, ed. New York: Braziller.

Pickett, S.T.A. 1976. Succession: an evolutionary interpretation. Am. Naturalist 110:107-119.

Pickett, S.T.A., and J.N. Thompson. 1978. Patch dynamics and the design of nature reserves. Biol. Conserv. 13:27-37.

Pickett, S.T.A., and P.S. White, eds. 1985. The ecology of natural disturbance and patch dynamics. Orlando, Fla.: Academic.

Pimm, S.L. 1991. The Balance of Nature? Ecological Issues in the Conservation of Species and Communities. Chicago: University of Chicago Press.

Pulliam, H.R. 1988. Sources, sinks, and population regulation. Am. Naturalist 132:652-661.

Pulliam, H.R., and B.J. Danielson. 1991. Sources, sinks, and habitat selection: A landscape perspective on population dynamics. Am. Naturalist 137:S50-S66.

Pulliam, H.R., J.B. Dunning, and J. Liu. 1992. Population dynamics in complex landscapes: A case study. Ecol. Appl. 2:165-177.

Risser, P.G. 1987. Landscape ecology: State of the art. Pp. 3-14 in Landscape Heterogeneity and Disturbance, M. G. Turner, ed. New York: Springer.

Rotenberry, J.T., and J.A. Wiens. 1980. Temporal variation in habitat structure and shrub steppe bird dynamics. Oecologia 47:1-9.

Schoener, T.W. 1986. Overview: Kinds of ecological Communities—Ecology becomes pluralistic. Pp. 467-479 in Community Ecology, J. Diamond and T.J. Case, eds. New York: Harper and Row.

Schroeder, R.L., and M.E. Keller. 1990. Setting objectives: A prerequisite of ecosystem management. Pp. 1-4 in Ecosystem Management: Rare Species and Significant Habitats. New York State Museum, Albany, N.Y.

Shafer, C.L. 1990. Nature Reserves: Island Theory and Conservation Practice. Washington, D.C.: Smithsonian Institution Press.

Shipley, W., and P.A. Keddy. 1987. The individualistic and community-unit concepts and falsifiable hypotheses. Vegetatio 69:47-55.

Society of American Foresters. 1993. Task force report on sustaining long-term forest health and productivity. Society of American Foresters, Bethesda, Md.

Soulé, M.E., ed. 1987. Viable Populations for Conservation. Cambridge, U.K.: Cambridge University Press.

Soulé, M.E., and B.A. Wilcox, editors. 1980. Conservation Biology: An Evolutionary-Ecological Perspective. Sunderland, Mass.: Sinauer Associates.

Stacey, P.B., and M. Taper. 1992. Environmental variation and persistence of small populations. Ecol. Appl. 2:18-29.

Thompson, J.N. 1982. Interaction and Coevolution. New York: John Wiley & Sons.

Turner, M.G., Y. Wu, L.L. Wallace, and W.H. Romme. 1994. Simulating winter interactions among ungulates, vegetation, and fire in northern Yellowstone Park. Ecol. Appl. 4:472-496.

van Horne, B. 1982. Population density as a misleading indicator of habitat quality. J. Wildl. Manage. 47:893-901.

Verner, J., K.S. McKelvey, B.R. Noon, R.J. Gutierrez, G.I. Gould, and J.W. Bock. 1992. The California Spotted Owl: A Technical Assessment of its Current Status. USDA Firest Service Report PSW GJR-133. U.S. Department of Agriculture, Forest Service, Washington, D.C.

von Bertelanffy, L. 1968. General System Theory: Foundations, Deverlopment, and Applications. Rev. Ed. New York: Braziller.

Wagner, F.H., and C.E. Kay. 1993. "Natural" or "healthy" ecosystems: Are U.S. national parks providing them. Pp. 257-270 in Humans as Components of Ecosystems: The Ecology of Subtle Human Effects and Populated Areas. New York: Springer.

Walker, B. 1989. Diversity and stability in ecosystem conservation. Pp. 121-130 in Conservation for the Twenty-First Century, D. Western and M.C. Pearl, eds. New York: Oxford University Press.

White, P.S. 1979. Pattern, process, and natural disturbance in vegetation. Bot. Rev. 45:229-299.

Whittaker, R.H. 1975. Communities and Ecosystems. 2nd Ed. New York: Macmillan.

Wiens, J.A. 1976. Population responses to patchy environments. Annu. Rev. Ecol. Syst. 7:81-120.

Wiens, J.A. 1984. On understanding a non-equilibrium world: myth and reality in community patterns and processes. Pp. 439-458 in Ecological Communities: Conceptual Issues and the Evidence, D.R. Strong, D. Simberloff, L. Abele and A.B. Thistle, eds. Princeton, N.J.: Princeton University Press.

6

Conservation Conflicts Between Species

Plants and animals are linked to other organisms in ecosystems in a variety of ways, so it is inevitable that conflicts will arise when attempts are made to protect individual species of plants or animals. Part of the charge presented to the Committee on Scientific Issues in the Endangered Species Act was to consider the severity of conflicting conservation needs when more than one species is listed[1] in the same geographical area and provide recommendations for resolving these conflicts.

INTERACTIONS OF SPECIES IN NATURE

To evaluate the potential problem of conservation conflicts between listed species, two fundamental ecological principles must be considered. The first principle is that organisms are components of networks in which they interact. This principle has significant implications in planning for survival and recovery of endangered species. If a management strategy does not account for the important relationships and interactions embodied in networks, then unexpected or untoward results can be expected (Holt and Talbot, 1978; Walker, 1989; Pickett et al. 1992; Fiedler et al., 1993; Franklin, 1993; Orians, 1993). For example, management of the New Jersey Pine

---

[1] Not all the species discussed in this chapter are listed under the ESA, but the cases illustrate the kinds of conservation conflicts that could occur between listed species.

Lands without consideration of certain rare herbs in the successions of pitch pine lowland or Atlantic white cedar swamp communities results in the decline in density of herb populations and their extirpation from many sites (Little, 1977; 1979). Likewise, managing Pennsylvania forests for high densities of white-tailed deer to support sporting interests inhibits tree regeneration in some stands and eliminates many shrub and herb species from the understory of the majority of forest stands (Marquis et al., 1975), although if hunting were prohibited, the problem would get worse.

Management or planning for recovery of endangered species in ignorance of the networks in which they exist is scientifically untenable. In situations where two or more endangered species are present, taking account of any network that includes them both (or all) should enhance the chances of optimizing the persistence and recovery of all species.

The second fundamental principle of ecology that is relevant to potential conflicts among listed species is that species are parts of spatial and temporal mosaics. The spatial dimension is cast in terms of mosaics because both natural and human-modified landscapes are conspicuously patchy (Pickett and White, 1985; Kolasa and Pickett, 1991; McDonnell and Pickett, 1993). This principle implies that the networks of interaction in which species exist have a spatial component.

The important ecological characteristics of mosaics for conservation of endangered species are that the resources, interactions, and constraints of endangered species can originate in the mosaic in components other than the current location of the listed entity (Risser, 1985). Although it is difficult to learn about the important ecological fluxes between patches that affect listed species, neglecting such fluxes can result in failures to preserve targeted species (Saunders et al., 1991; Tyser and Whorley, 1992).

Taken together, the two fundamental principles described above suggest improvement in endangered species management. Management that views each species as an entity by itself, with no or little attention to the network of interactions, is likely to produce faulty protection strategies. Likewise, neglecting the larger spatial context in which species exist may well miss key forces that are needed to maintain the species. Taking these two principles into account in planning for management of listed species can provide a basis for assessing the potential for species conflicts and mitigating them effectively. Without considering networks and spatial contexts, species conflicts are relegated to ad hoc solutions on a crisis footing.

The committee has found few well-documented cases in which management practices focusing on particular species protected under the Endangered Species Act have resulted in direct conflict between conservation needs for the species. Such situations likely will increase, however, as more species are listed and as species and their networks become better understood. A sample of a few specific cases demonstrates how management practices directed solely at single species might present such problems.

NORTHERN GOSHAWK AND MEXICAN SPOTTED OWL

Potential conflicts arise between the northern goshawk and the Mexican spotted owl, two species of avian predators, when the U.S. Forest Service attempts to manage habitat for one or the other in areas where both species occur. In the case of the Mexican spotted owl, current management favors comparatively large, dense stands of closed canopy forest. However, guidelines for the northern goshawk require a much more varied forest condition, with major areas of low stand density, open canopies, and numerous forest openings. In addition, the Fish and Wildlife Service postulates that with more openings in the forest (i.e., forest fragmentation), the Mexican spotted owl will be increasingly susceptible to avian predation (*Fed. Reg.* 58:14269). Thus, such habitat manipulation might increase direct loss to the listed species through predation by the northern goshawk (see Box 6-1).

WINTER-RUN CHINOOK SALMON AND DELTA SMELT

Another example involves the endangered winter-run chinook salmon and the threatened delta smelt, two species of fish that occur where the Sacramento and San Joaquin rivers meet in the Central Valley region of California (see Box 6-2). The 1992 Central Valley Improvement Act (Title 34 of P. L. 102-575) contains a series of provisions that dictate water use and contracting for the Central Valley Project, as well as mitigation and restoration activities that will benefit the threatened fish species in the area. Revenues from water users and other direct beneficiaries of the water project are to pay for the restoration costs.

Some of the provisions call for measures that will benefit both the winter-run salmon and the delta smelt. For example, Provision 4 calls for the improvement of screens and fish-recovery facilities at the Tracy pumping plant. Benefits to the delta smelt will accrue because of reduced entrainment (destruction of fish or larvae at the intake mechanisms of diversion facilities such as those in the delta (*Fed. Reg.* 59:816)). These measures will also protect the stronger-swimming salmon. Likewise, Provision 7 states that the Central Valley Project must comply with all applicable flow standards that apply to it, including any new regulations that might be imposed. Such regulations should benefit both species. Furthermore, the recent federal listing of water-quality standards for the Sacramento and San Joaquin rivers, the San Francisco Bay, and the delta region (*Fed. Reg.* 59:810-852) and the recent critical habitat listing for the delta smelt (*Fed.*

Box 6-1
Northern goshawk and Mexican spotted owl

The northern goshawk, the largest member of the genus *Accipiter*, is a forest habitat generalist that uses a variety of forest types, forest ages, structural conditions, and successional stages (Reynolds et al., 1992). The northern goshawk is listed as a "sensitive species" by the Southwest Region of the Forest Service because of concerns that its populations and reproduction may be declining in this region due in part to forest changes caused by historic timber harvest patterns.

In response to concern for this species, the Forest Service has developed and adopted a set of management recommendations for the northern goshawk in the southwestern United States (Reynolds et al., 1992). However, rather than focusing solely on the goshawk, these recommendations are designed to provide habitat for many of the goshawk's prey species, such as the American robin (*Turdus migratorius*), band-tailed pigeon (*Columba fasciata*), mourning dove (*Zenaida macroura*), blue grouse (*Dendragapus obscurus*), hairy woodpecker (*Picoides villosus*), northern flicker (*Colaptes auratus*), red-naped sapsucker (*Sphyrapicus nuchalis*), Williamson's sapsucker (*Sphyrapicus thyroideus*), Steller's jay (*Cyanocitta stelleri*), chipmunks (*Tamias* spp.), golden-mantled ground squirrel (*Citellus lateralis*), red squirrel (*Tamiasciurus hudsonicus*), tassel-eared squirrel (*Sciurus aberti*), and cottontails (*Sylvilagus* spp.). Management for these species will result in a quite varied forest condition, with openings and low density forest occurring fairly commonly.

The Mexican spotted owl (*Strix occidentalis lucida*) was listed as a threatened species on March 16, 1993 (*Fed. Reg.* 58:14248-14271). This medium-sized owl is found in central Colorado and Utah south through Arizona, New Mexico, and western Texas, primarily in canyons and areas with steep slopes. It is believed to be threatened owing to loss and modification of its forest habitat due to timber harvest and fire and increased predation due to habitat fragmentation. When found in forested habitats, the Mexican spotted owl is believed to prefer areas with high canopy closure, high stand density, and a multilayered canopy for its nesting, roosting, and foraging sites (Ganey et al., 1988; Ganey and Balda, 1989; Fletcher, 1990). Great horned owls (*Bubo virginianus*) and red-tailed hawks (*Buteo jamaicensis*) have been identified as possible predators, and goshawks as probable predators, of the Mexican spotted owl (Skaggs, 1990).

Reg. 59:852-861) attempt to restore the important areas of the drainage system to historic salinity levels.

In one case, however, a provision of the Central Valley Improvement Act might benefit one of the threatened species and have an adverse effect on the other. Provision 14 calls for modification of the flow and control structures at the Delta Cross Channel; it is primarily for the benefit of striped bass, a popular nonnative sport fish. This provision might benefit the winter-run salmon, because it will reduce the amount of water entering

---

Box 6-2
Chinook salmon and delta smelt

Four races of chinook salmon (*Oncorhynchus tshawytscha*) inhabit the Sacramento River system. The winter-run race migrates upstream in the winter. Although it was previously abundant in this drainage system, population sizes have declined drastically, from an estimated 117,808 individuals in 1969 to 341 in 1992/93 (FWS, 1992). This race is now listed as endangered on the Federal Endangered Species List (*Fed. Reg.* 59:13836). Adults migrate as far up the Sacramento River as possible in the winter, with spawning as early as mid-April, reaching a peak in June, and then declining through the summer until August. Eggs are incubated for 40-60 days followed by an additional 2-4 weeks for the newly hatched fry on gravel substrate. Incubation must occur in cool water temperatures (43-58°F), although incubation occurs during the hottest time of the year. Migration of the juveniles begins after a short period of growth, with juveniles migrating to the lower river up to a year after the beginning of spawning of their cohort.

The delta smelt (*Hypomesus transpacificus*) is endemic to the Sacramento-San Joaquin River estuary. The entire species is listed as threatened on the Federal Endangered Species List (*Fed. Reg.* 58:12863). Compared with salmon, the smelt has a short life span. Spawning occurs primarily from December to March. Eggs hatch 10-12 days after fertilization into larvae that drift downstream with the river current. By one year of age the fish are sexually mature, with mating of a cohort continuing into the summer of the year after hatching. Although the biology of this species is not known in detail, it appears that adults die soon after spawning. The delta smelt is associated with well-oxygenated, very cold water. It also appears that hard substrates and submerged rocks are needed for successful spawning

Two major water projects, the Central Valley Project (CVP) and the State Water Project (SWP), and many smaller diversions affect these species both in the Sacramento-San Joaquin delta and the Sacramento River (for the winter-run chinook salmon). These water diversions can entrain fish along the diverted flows as well as reduce flows downstream. Flow diversions and impoundment storage behind dams can greatly alter flows, flow pattern, and seasonality. Flows also affect movement of fish, particularly larvae of delta smelt and striped bass. In addition, flows play a major role in the location of highly productive areas for phytoplankton and zooplankton. An "entrapment" or "null" zone that provides important nursery habitat for delta smelt and striped bass is typically formed in Suisun Bay downstream of the delta. During drought years, this zone occurs in the channel of the delta much closer to the CVP and SWP water intakes than in normal years. During such periods, entrainment is expected to be increased. In addition, production of prey organisms is expected to decrease because of the smaller size of the delta channels.

An additional problem associated with the major water projects is increased predation by fish-eating predators, including adult striped bass, which use features of the major intakes to prey on smaller fish. Such predation is considered to be one of the major sources of loss associated with the SWP.

the central delta while increasing the San Joaquin and Mokelumne rivers flows to the pumps. Delta smelt will be adversely affected, however, because they will be in these waters at the times of reduced flows, and entrainment at the Tracy pumping plant might increase. Moreover, striped bass are probably predators of the smelt, so this action could result in increased mortality of the smelt. Another provision, Provision 18, advocates restoration of the striped bass fishery, which would also have an adverse effect on the delta smelt due to increased predation (see Box 6-2).

It is apparent that the Central Valley Improvement Act attempts to correct many problems that the Central Valley Project causes for threatened and endangered fish and that, in some cases, actions will benefit all of the species involved. However, in some situations, management techniques that are most beneficial to one threatened species adversely affect the other. In this system, the situation is complicated by the presence of a third species, striped bass, that is being considered because of public interest. Tradeoffs will have to be evaluated in each case to determine what measures should be implemented. For example, additional flow releases without additional pumping after delta smelt spawning could benefit that species by carrying eggs and larvae past the pumps to Suisun Bay where the best rearing habitats occur during normal flow years.

Other actions affecting these species include the water-quality standards set by the U.S.Environmental Protection Agency to protect the delta. Those standards (*Fed. Reg.* 59:810-852) will result in increased flows and decreased pumpings, which should help normalize salt levels and provide larger quantities of water to facilitate migration conditions and rearing. In addition, the State Water Resource Control Board is under federal court mandate to impose standards and regulations as well. Such judicial and regulatory actions have resulted in pumping restrictions at the State Water Project and Central Valley Project to reduce losses of winter-run salmon and delta smelt. Nevertheless, it is unclear whether these actions will be able to help both species, or if one will still suffer at the expense of the other.

BACHMAN'S SPARROW AND RED-COCKADED WOODPECKER

Management decisions designed to improve conditions for a threatened or endangered species may inadvertently affect dozens of nontarget species found in the same habitats. Possible effects on nontarget species are rarely assessed before implementation of management actions. It will become increasingly important to develop tools to assess the effect of proposed management strategies on a wide variety of organisms as federal agencies and others put increased emphasis on management for biodiversity.

Several research groups are developing population simulation models linked to Geographical Information System (GIS) maps that capture some of the complexity of real-world landscapes and allow simultaneous consideration of the responses of many different species to management proposals. This approach requires, at a minimum, a good understanding of the habitat requirements of the species of interest, and a realistic landscape map that shows the locations of current and future suitable habitat as a function of management decisions. More recent versions of these models also require detailed information on habitat-specific demography and dispersal behavior.

Liu (1992) and Liu et al. (1995) give one example of the use of such models to forecast how management plans largely designed for one endangered species might affect a nontarget species. Liu et al. (1995) used a mobile animal populations model (Pulliam et al., 1992) (Chapter 5) to determine how the Bachman's sparrow (*Aimophila aestivalis*), a declining species of management interest in the southeastern United States, might respond to a management strategy largely designed to favor populations of the endangered red-cockaded woodpecker (*Picoides borealis*) (see Box 6-3).

The results of such models are useful in a variety of ways. In the particular case discussed, the model allowed alternative cutting and thinning plans to be explored, at least some of which allowed larger populations of sparrows, as well as woodpeckers, to be maintained. Models of this sort can also incorporate economic considerations (Angelstam, 1992; Liu et al., 1994) and might prove useful in future attempts to balance ecological and economic goals. However, caution must be exercised in using models of this sort for management decisions, because the models are not yet fully quantified or tested against field results. A prudent use of such models would be in the context of adaptive ecosystem management. Here, the models would be used to generate testable hypotheses, and forest-management practices would be used as an experiment to test the model predictions.

MARINE MAMMALS AND SALMONIDS

The effect of predation by marine mammals on salmonids has been controversial since at least the 19th century (Merriam, 1901). In 1899, the president of the California Board of Fish Commissioners proposed to kill "10,000 of the 30,000 [California sea lions, *Zalophus californianus,* and Steller's sea lions, *Eumatopias stelleri*] that now infest our harbor entrances and contiguous territories" to reduce their alleged depredations on salmon. Merriam pointed out that there probably weren't even 10,000 sea lions on the coast. He described the work of L.L. Dyche, who examined the stom-

---

Box 6-3
Bachman's sparrow and red-cockaded woodpecker

Although formerly much more widespread, Bachman's sparrow populations have disappeared from much of the historic range and are now restricted mostly to the southeastern coastal plain, where they can be found both in mature pine forest and in some early successional habitats. The species is found in habitats with a dense ground cover of grasses and forbs and relatively open understory with few shrubs (Dunning and Watts, 1990). Mature pine forests (over 80 years old) managed for the red-cockaded woodpecker usually provide adequate habitat for Bachman's sparrow, particularly if the understory is burned periodically. The species is also relatively common in the young (1-5 years old) successional pine stands that follow clearcutting. Intermediate-age pine stands (approximately 6-80 years old) are not suitable for Bachman's sparrow, presumably because the relatively closed canopy prevents a dense ground layer vegetation from forming.

Liu et al. (1994) developed a MAP model to study Bachman's sparrow responses to current and proposed forest management plans on the Savannah River Site (SRS), a large region of pine forest in South Carolina managed by the U.S. Forest Service for timber production and wildlife conservation. The Forest Service has developed a 50-year operations plan for the SRS that considers the habitat requirements of over 42 plant and animal species of management concern. However, most of the specific management practices described in the operation plan are aimed at improving habitat for the endangered red-cockaded woodpecker. Using the operation plan, Liu and coworkers projected future habitat conditions at SRS to determine how the proposed management practices would impact the Bachman's sparrow, a species that is not a target of most of the specified management strategies. According to these researchers, the forest-management practices proposed in the operation plan would have a strong effect on the Bachman's sparrow. The model simulations suggested that, in the long run, the sparrow would benefit from the changes because, under the plan, the acreage of mature pine forest of the sort suitable for the red-cockaded woodpecker and Bachman's sparrow would increase substantially. However, the simulations suggest that the sparrow might decline precipitously during the first few decades of operation of the plan because of decline in the availability of early successional habitat.

---

achs of 25 sea lions and found mainly octopus and squid, but no fish, and he argued against killing sea lions. Bonnot (1928) described the killing of sea lions in great detail; they were hunted in Oregon for bounties, in California for their penises and testicles (known as "trimmings") and for various other purposes along the coasts, including to prevent them from eating fish. Adult sea lions were killed with guns and landmines, and pups by drowning them in weighted sacks (Bonnot, 1928).

Fish predation by sea lions has received attention again recently, largely because of its high visibility in one location: the Hiram M. Chittenden locks in Ballard (a section of Seattle). The Chittenden (or Ballard) locks were completed in 1917 as part of a project of the U.S. Army Corps of Engineers to allow ship traffic between Puget Sound and Lake Washington (Willingham, 1992). Previously, the lake had drained through the Black River, which flowed into the Cedar and Green rivers and then into Puget Sound. The current drainage into Puget Sound is through the Lake Washington Ship Canal, Lake Union, and the Ballard locks. Steelhead trout (*Oncorhynchus mykiss*) migrated through Lake Washington to spawn in its tributaries before 1917; they do so today. The runs consist of wild and hatchery fish, but the hatchery program has been discontinued recently (Fraker, 1994). Although some people believed the current runs were derived from hatchery fish, genetic analysis indicates that they are probably descendants of the original wild runs (Fraker, 1994).

Lake Washington steelhead are winter-run fish, returning to the fresh water to spawn from December to April. As the fish enter the ship canal below the locks, some of them are captured and eaten by California sea lions. The first observation of such predation was made in 1980; by the mid-1980s there were as many as 60 sea lions in the area around the locks and more than 50% of the returning steelhead were being eaten. (Between 51% and 65% were taken each year up to 1992, except for 1985-1986 (15%) and 1986-1987 (41%), when predator-control efforts had some success.) The number of fish in the run that escaped to spawn, which ranged from 474 to 2,575 fish from 1980-1981 to 1985-1986, declined to 184 fish in 1992-1993 (Fraker, 1994) and to 70 in 1993-1994 (NMFS and WDFW, 1995).

It is clear that sea lions have affected the Lake Washington steelhead run, but they are not entirely responsible for its recent decline. Cooper and Johnson (1992) found that steelhead had declined generally since 1985. They considered the following items to be possible contributing factors: competition for food with other salmon, in particular, 8 billion hatchery salmon released since the late 1980s; authorized and unauthorized drift-net fisheries (probably not currently a factor); predation by birds and mammals; and large-scale environmental changes.

Predation by marine mammals is probably not a major factor in the current decline of salmon in general. Anadromous salmonids and marine mammals coexisted for thousands of years before the current declines in salmonids, and California and Steller sea lions were much more abundant in the first half of the 19th century—a time when salmon were also abundant—than later. And marine mammals do not normally specialize on salmonids; they eat a wide variety of prey items, determined by what is available and how easy it is to catch (Gearin et al., 1988; Fraker, 1994; Olesiuk, 1993). Finally, the Ballard locks area provides a local concentration of fish in space and time, and they have few refuges there, and sea lions congregate there in large numbers (Fraker, 1994).

However, many marine-mammal populations are increasing, at least partly because of the protections of the Marine Mammal Protection Act (MMPA). California sea lions now number more than 100,000 (Fraker, 1994). Other human activities, combined with increasing marine-mammal populations, could cause increasing problems, especially in areas where the fish congregate, as in the case of the steelhead at Ballard. If the Lake Washington steelhead were listed as endangered or threatened under the Endangered Species Act, the conflict would be brought into sharper focus by the requirements of the Endangered Species Act and the MMPA. Indeed, the MMPA was amended in 1994 to allow the killing of marine mammals under particular circumstances, and Washington state has filed a petition to remove sea lions from the Ballard locks and kill them if all other methods to keep them from eating steelhead fail (NMFS and WDFW, 1995).

CONCLUSIONS

We have been able to document only these and a few other cases of conflicting conservation needs resulting from management plans targeted toward individual species. It is possible that this low number stems from several factors: lack of knowledge of the networks of which threatened and endangered species are part; the fact that comparatively few species are currently listed and that recovery plans for even fewer have been formulated; and the inadvertent protection for other listed species under some current recovery plans. We expect, however, that the potential for such conflicts will rise as ecologies of listed species become better known, more recovery plans are formulated, and habitat for conserving endangered species becomes more constricted.

The greatest potential for conflicts in protecting species and for management of individual species under current policies will arise in situations in which habitat reductions—especially extreme reductions—themselves are the causes of endangerment and the habitats of listed species are largely overlapping. Resolution of such conflicts will have to be made on a case by case basis. A process should be devised that will facilitate such resolutions using analyses of risk and recovery as outlined in Chapter 8.

The most effective way to avoid conflicts resulting from individual management plans is to maintain large enough protected areas for listed species to allow the existence of mosaics of habitats and dynamic processes of change within these areas. In addition to and as part of this strategy, multispecies plans should be devised that ensure the maintenance of habitat mosaics and ecological networks. Habitat (in the broadest sense) thus plays a crucial role in protecting individual target species and, ultimately, in reducing the need for listing additional species.

The Fish and Wildlife Service has prepared a number of packages listing multiple species in the same ecosystems, and it has agreed in a recent judicial settlement "to direct each region, where biologically appropriate, to use a multi-species, ecosystem approach to their listing responsibilities under the ESA" (settlement agreement, The Fund for Animals v. Lujan, Civ. No. 92-800, December 15, 1992). This is an important directive whose implementation and oversight warrant priority. Key questions should be addressed as the initiative moves forward: Are listing resources best deployed to advance the policy? To what degree have staff and consulting resources been arranged to optimize cooperative work across taxa? To what extent have the FWS and National Marine Fisheries Service made a special effort to identify widespread species presenting great potential for conflict, but also for conflict resolution? Finally, should the National Biological Service (NBS) be an important vehicle for ensuring that these questions can be answered in the affirmative?

RECOMMENDATIONS

Because of the interactions of plants and animals with other organisms in their environments, the most effective way to avoid conflicts resulting from individual management plans of co-occurring endangered species is to maintain large enough protected areas for listed species to ensure the presence of habitat mosaics and to allow for the dynamic processes of change that will inevitably occur within such areas. As part of this strategy, multispecies plans (e.g., habitat conservation plans; see Chapter 4) should be devised that ensure habitat mosaics and ecological networks are maintained.

When the available habitat is insufficient to avoid conflicts, the analysis of options will have to be done separately for each situation. In most cases, long-term results are more important than short-term ones. In the example of Bachman's sparrow and the red-cockaded woodpecker (Box 6-3), both birds would benefit in the long term under the Forest Service's plan, despite short-term declines in the sparrow's population. Other considerations would be which species is most likely to suffer irreversible harm if its needs are not fully addressed, the taxonomic level of the populations involved (e.g., a full species is probably more important than a distinct population segment), and ecological considerations (e.g., would the loss of one species have a greater effect on the ecosystem than the loss of the other?).

REFERENCES

Angelstam, P. 1992. Conservation of communities—The importance of edges, surroundings and landscape mosaic structure. Pp. 9-70 in Ecological Principles of Nature Conservation: Applications in Temperate and Boreal Environments, L. Hansson, ed. New York: Elsevier Applied Science.

Bonnot, P. 1928. Report on the seals and sea lions of California. Fish Bulletin 14. California Division of Fish and Game, Sacramento, Calif.

Cooper, R., and T.H. Johnson. 1992. Trends in steelhead (*Oncorhynchus mykiss*) abundance in Washington and along the coast of North America. Washington Department of Wildlife Report 92-20.

Dunning, J.B., and B.D. Watts. 1990. Regional differences in habitat occupancy by Bachman's sparrow. Auk 107:463-372.

Fiedler, P.L., R.A. Laidy, R.D. Laven, N. Gershenz, and L. Saul. 1993. The contemporary paradigm in ecology and its implications for endangered species conservation. Endangered Species Update 10:7-12.

Fletcher, K.W. 1990. Habitats Used, Abundance and Distribution of Mexican Spotted Owl (*Strix occidentalis lucida*) on National Forest System Lands. U.S. Department of Agriculture Forest Service, Southwest Region, Albuquerque, N.M. 55 pp.

Fraker, M. 1994. California Sea Lions and Steelhead Trout at the Chittenden Locks, Seattle, Washington. Marine Mammal Commission, Washington, D.C.

Franklin, J. F. 1993. Preserving biodiversity: Species, ecosystems, or landscapes. Ecol. Appl. 3:202-205.

FWS (U.S. Fish and Wildlife Service). 1992. Measures to Improve the Protection of Chinook Salmon in the Sacramento/San Joaquin River Delta. WRINT-USFWS-7. Expert testimony of U.S. Fish and Wildlife Service on chinook salmon technical information for State Water Resources Control Board Water Rights Phase of the Bay/Delta Proceedings, July 6.

Ganey, J.L., and R.P. Balda. 1989. Distribution and habitat use of Mexican spotted owls in Arizona. Condor 91:355-361.

Ganey, J.L., J.A. Johnson, R.P. Balda, and R.W. Skaggs. 1988. Status report: Mexican spotted owl. Pp. 145-150 in Proceedings of the Southwest Raptor Management Symposium and Workshop, R.L. Glinski, B.G. Pendleton, M.B. Moss, M.N. LeFranc, Jr., B.A. Milsap, and S.W. Hoffman, eds. National Wildlife Federation, Washington, D.C.

Gearin, P., R. Pfeifer, S.J.J. Jeffries, R.L. DeLong, and M.A. Johnson. 1988. Results of the 1986-1987 California Sea Lion-steelhead Trout Predation Control Program at the Hiram M. Chittenden Locks. Northwest and Alaska Fisheries Center Processed Report 88-30. Alaska Fisheries Science Center, Seattle, Wash.

Holt, S.J., and L.M. Talbot. 1978. New principles for the conservation of wild living resources, Vol. 59. Wildlife Society, Louisville, Ky.

Kolasa, J., and S.T.A. Pickett, eds. 1991. Ecological Heterogeneity. New York: Springer.

Little, S. 1977. Wildflowers of the pine barrens and their niche requirements. N.J. Outdoors 1:16-18.

Little, S. 1979. Fire and plant succession in the New Jersey pine barrens. Pp. 297-314 in Pine Barrens: Ecosystem and Landscape, R.T.T. Forman, ed. New York: Academic.

Liu, J. 1992. ECOLECON: A Spatially Explicit Model for Ecological Economics of Species Conservation in Complex Forest Landscapes. PhD Dissertation. University of Georgia.

Liu, J., F. Cubbage, and H.R. Pulliam. 1994. Ecological and economic effects of forest landscape structure and rotation length: Simulation studies using ECOLECON. Ecol. Econ. 10:249-263.

Liu, J., J.B. Dunning, Jr., and H. Ronald Pulliam. 1995. Potential effects of a forest management plan on Bachman's Sparrows (*Aimophila aestivalis*): Linking a spatially explicit model with GIS. Conserv. Biol. 9(1):62-75.

Marquis, D.A., T.J. Grisez, J.C. Bjorkbom, and B.A. Roach. 1975. Interim Guide to Regeneration of Alleghany Hardwoods. USDA Forest Service Gen. Tech. Rep. NE-19. U.S.

Department of Agriculture Forest Service, Northeastern Forest Experiment Station: Upper Darby, PA.

McDonnell, M.J., and S.T.A. Pickett, eds. 1993. Humans as Components of Ecosystems: The Ecology of Subtle Human Effects and Populated Areas. New York: Springer.

Merriam, C.H. 1901. Food of sea lions. Science N.S. 13:777-779.

NMFS and WDWF (National Marine Fisheries Service and Washington Department of Fish and Wildlife). 1995. Environmental Assessment on Protecting Winter-Run Steelhead from Predation by California Sea Lions in the Lake Washington Ship Canal, Seattle, Wash. National Marine Fisheries Service Northwest Regional Office, Seattle, and Washington Department of Fish and Wildlife, Olympia.

Olesiuk, P.F. 1993. Annual prey consumption by harbor seals (*Phoca vitulina*) in the Strait of Georgia, British Columbia. Fish. Bull. 91:491-515.

Orians, G.H. 1993. Endangered at what level? Ecol. Appl. 3:206-208.

Pickett, S.T.A., V.T. Parker, and P. Fiedler. 1992. The new paradigm in ecology: Implications for conservation biology above the species level. Pp. 65-88 in Conservation Biology: The Theory and Practice of Nature Conservation, Preservation, and Management, P. Fiedler and S. Jain, eds. New York: Chapman and Hall.

Pickett, S.T.A., and P.S. White, eds. 1985. The Ecology of Natural Disturbance and Patch Dynamics. Orlando, Fla.: Academic.

Pulliam, H.R., J.B. Dunning, and J. Liu. 1992. Population dynamics in complex landscapes: a case study. Ecol. Appl. 2:165-177.

Reynolds, R.T., R.T. Graham, M. Hildegard, R.L. Bassett, P.L. Kennedy, D.A. Boyce, Jr., G. Goodwin, and E.L. Fisher. 1992. Management Recommendations for the Northern Goshawk in the Southwestern United States. Fort Collins, CO: Rocky Mountain Forest and Range Experiment Station; Albuquerque, NM: Southwestern Region, Forest Service, U.S. Department of Agriculture. USDA Gen. Tech. Rep. RM-217. 90 pp.

Risser, P.G. 1985. Toward a holistic management perspective. BioScience 35:414-418.

Saunders, D.A., R.J. Hobbs, and C.R. Margules. 1991. Biological consequences of ecosystem fragmentation: A review. Conserv. Biol. 5:18-32.

Skaggs, R.W. 1990. Spotted owl telemetry studies in the Lincoln National Forest, Sacramento Division: Progress Report. New Mexico Department of Game and Fish, Santa Fe. 7 pp.

Tyser, R.W., and C.A. Whorley. 1992. Alien flora in grasslands adjacent to road and trail corridors in Glacier National Park, Montana (USA). Conserv. Biol. 6:253-262.

Walker, B. 1989. Diversity and stability in ecosystem conservation. Pp. 121-130 in Conservation for the Twenty-first Century, D. Western and M.C. Pearl, eds. New York: Oxford University Press.

Willingham, W.F. 1992. Northwest Passages: A History of the Seattle District U.S. Army Corps of Engineers, 1896-1920. U.S. Army Corps of Engineers, Seattle District.

7

Estimating Risk

The concept of risk is central to the implementation of the Endangered Species Act. The committee was asked to review the role of risk in decisions made under the act, review whether different levels of risk apply to different types of decisions made under the act, and identify practical methods for assessing risk.

Risk is the probability that something (usually a bad outcome) will occur. *Risk assessment* aims to estimate the likelihood of a particular (usually bad) outcome occurring. *Risk management* is an integrating framework that assesses the likelihood of bad outcomes and analyzes ways to minimize the risk of bad outcomes, or at least to respond appropriately if they occur. Many risk assessments follow the framework developed by the National Research Council to apply to human health (NRC, 1983); an example of a specific risk assessment framework is the one developed by EPA (Risk Assessment Forum, 1992), which tracks patterns of exposure to harmful substances and responses of ecological systems to these exposures. The sometimes confusing terminology of risk assessment and some of the issues in applying risk assessment to ecological systems were described by Policansky (1993); further examples were discussed by the National Research Council (NRC, 1993).

The main risks involved in the implementation of the Endangered Species Act are the risk of extinction and the risks associated with unnecessary expenditures or curtailment of land use in the face of substantial uncertainties about the accuracy of estimated risks of extinction and about future events. In this chapter, we consider the problem of estimating the risk of extinction and the limitations of our current ability to estimate this risk. Models are an important tool for analyzing the consequences of complex processes, because intuition is often not reliable. In some cases, the predictions of the models discussed are not precise because information is lacking or because the underlying processes are not fully understood. They are valuable as guides to research and as tools for analyzing the comparative effects of various environmental and management scenarios.

ESTIMATING THE RISK OF EXTINCTION

Since the inception of the ESA, there have been enough developments in conservation biology, population genetics, and ecological theory that substantial scientific input can be used in the listing and recovery-planning processes. The following synthesizes and evaluates the various approaches and conclusions that have emerged from recent attempts to understand the vulnerability of small populations to extinction. The material focuses on random changes in population sizes and in their structure, changes in genetic variability, environmental fluctuations, and habitat fragmentation. Additional theoretical and field research are needed to resolve or reduce uncertainties, but existing analyses give insight into the relative magnitude and possible scaling of various influential factors in the extinction process. More thorough and technical reviews were provided by Dennis et al. (1991), Thompson (1991), and Burgman et al. (1992).

SOURCES OF RISK

Habitat loss, effects of introduced species, and in some cases overharvesting are almost always the ultimate causes of species extinction. Decline of populations to a low density makes them vulnerable to chance events and sets into play the extinction risks outlined below. When conditions have deteriorated to the point that a wild population cannot maintain a positive growth rate, no sophisticated risk analysis is required to tell us that extinction is inevitable without human intervention. Our attention will be focused instead on cases in which a population with a positive capacity for growth in an average year is still vulnerable to chance events that cause short-term excursions to low densities. Limitations of these approaches are discussed at the end of the section.

Random Demographic Changes

Demographic features, such as family size, sex, and age at mortality, vary naturally among individuals. In populations containing more than about 100 individuals, individual variation averages out and has little effect on the dynamics of population growth. However, in small populations, random variation in demographic factors can occasionally reach such an extreme state that extinction is certain. This can arise, for example, if all members of one sex die before reaching maturity or if all progeny are of the same sex, as was the case in the dusky seaside sparrow (*Ammodramus maritimus nigrescens*) after loss of habitat led to its population decline.

Substantial effort has been expended to develop general models for predicting the risk to small populations of extinction due to demographic stochasticity. Several assumptions must be made about the ways in which populations grow, in particular, about the way population growth rates respond to density. From the standpoint of an endangered species, the simplest conceivable model assumes that the population has been pushed to its limits—resources (habitat and food availability) have become so scarce that, on average, the expected number of births in an interval is the same as the expected number of deaths. In this case, with individual births and deaths being random, the mean time to extinction for a population starting with N individuals is simply N generations (Leigh, 1981), i.e., the time to extinction increases linearly with the population size. (Box 7-1 contains definitions of terms; Box 7-2 has definitions of symbols used in analyses.)

A more common situation is one in which resources are sufficient to support an average positive population growth when the population density is below a threshold. Due to chance, the actual growth rate in any generation will deviate somewhat from its expected value, and in the rare event that the cumulative growth rate realized over several consecutive generations is sufficiently negative, the population size will be reduced to zero (i.e., extinction will occur).[1]

All the demographic models discussed in this section assume that all members of the population are functionally identical. There is no variation based on age or sex; individuals are assumed to be identical with respect to reproductive and mortality rates. Thus, strictly speaking, the results apply best to short-lived asexual organisms or to hermaphrodites that synchronously reproduce toward the end of their life, as do many annual plants and some invertebrates. Models incorporating age structure, which are appropriate for vertebrates, require information on the mean and variance of age-specific mortality and fecundity schedules (Lande and Orzack, 1988; Tuljapurkar, 1989), information that is limited for even the best-studied species in nature.

For species with separate sexes (most vertebrates and many other or-

---

[1]With this type of model, the mean time to extinction increases exponentially with the product of the expected population growth rate at low density, $\bar{r}$, and the population carrying capacity, K, where K can be viewed as the number of individuals that a reserve can sustain at stable density (see Example 7-1).

Box 7-1
Definition of Terms

adaptive variation—genetic variation for characters upon which natural selection operates and which may be favored within the range of environments experienced by a species.

character—the overt (phenotypic) expression of a gene or group of genes.

deme—a local population of interbreeding organisms.

density dependence—the influence of population density on a specific phenomenon, e.g., density-dependent growth.

effective population size—the number of breeding adults that would give rise to the rate of inbreeding observed in a population if mating were at random and the sexes were equal in number. The effective population size is always less than the actual population size.

fitness—relative reproductive success or genetic contribution to future generations.

gametic mutation rate—the average total number of new mutations arising *de novo* in a gamete.

gene pool—the total set of genes contained within a population or species.

genetic variance—variability in the genomes of individuals within and between populations.

genome—the complete set of genetic material carried by an individual.

genotype—the specific set of genes—including the specification of their allelic forms—carried by an individual; may refer to a single genetic locus (e.g., blood genotype of an individual) or to the allelic forms of the complex of genes influencing the expression of a multifactorial trait.

homozygous—most species inherit parallel sets of genes from their parents. For a gene with a particular function, a homozygous individual is one that inherits identical copies of the gene (i.e., the same allelic form) from both parents. If the allelic forms are different, the individual is **heterozygous.**

mutation—a heritable change in a gene.

outbreeding depression—a reduction in fitness in the hybrid progeny, or later descendants, of crosses between members of different populations.

population—a group of closely related, interbreeding individuals.

population bottleneck—a transient and extreme reduction in population size relative to normal population sizes such that genetic diversity is reduced simply by the reduction in population size.

random genetic drift—changes in gene frequencies arising from chance sampling of gametes in small populations.

stochasticity—random variation.

Box 7-2
Definitions of mathematical notation

C the rate at which a subpopulation will recolonize an area.

E probability of extinction of a subpopulation.

K population carrying capacity; number of adults the environment can support.

N population size.

N_e effective population size.

p_e the average probability of extinction per generation. Only in the special case that p_e is constant in time does $\bar{t}_e = 1/p_e$.

$\bar{r}$ the intrinsic rate of population growth; i.e., the expected exponential rate of population increase at densities less than K.

$\bar{r}$ the long-run average growth rate; equal to $\bar{r} - V_e/2$.

s the selection intensity operating against a deleterious mutation in the homozygous state. For example, if the deleterious gene affects viability to maturity such that $s = 0.05$, then a homozygote for the deleterious allele (all other things being equal) has a 5% reduction in the probability of surviving to maturity.

$\bar{t}_e$ the mean time to extinction, measured in generations.

V_e between-generation variance of the population growth rate; i.e., the mean squared deviation of $\bar{r}$ in any generation from the expected value of $\bar{r}$.

μ the genomic deleterious mutation rate. Almost no data exist on this, except for *Drosophila*, although a fair amount of empirical work is now going on to fill this gap in our knowledge. The general principle of all experiments to estimate μ is the same—start with a genetically uniform stock; create sublines; maintain the sublines in isolation from each other with a minimum possible population size (to minimize the efficiency of natural selection against new mutations); and then over time watch the lines decline and diverge in terms of mean fitness. The details are somewhat complex statistically, but from this information (on the rate of decline in mean fitness and the rate of divergence of subline-specific fitness), it is possible to get a downwardly biased estimate of μ (Mukai, 1979; Houle et al., 1992).

Example 7-1

Suppose that the expected per capita growth rate is an average $\bar{r}$ so long as the population size is below a carrying capacity K defined by the features of the environment. ($\bar{r}$ is a logarithmic growth rate, and an $\bar{r}$ equal to 0.7/year implies that, at low density, the population can double its size in a year.) Assuming the population actually has positive growth potential ($\bar{r} > 0$), the mean extinction time (in generations) is approximately $(1 + \bar{r})e^{2\bar{r}K}/(2K\bar{r}^2)$, where $e = 2.72$ is the base of natural logarithms (Leigh, 1981; Goodman, 1987a,b). Notice that this expression can be written as $((1 + r)/r)(e^{2rK}/2rK)$. The term $(e^{2rK}/2rK)$ entirely determines the scaling with the population carrying capacity; i.e., the scaling with K depends on the composite parameter **rK**. The term $(1 + r)/r$ is independent of population size. This shows that if random fluctuations in offspring number and individual survivorship alone were responsible for extinction, the mean extinction time would scale nearly exponentially with the product of the rate of increase and the carrying capacity of the environment. Unless this product is very small, this model predicts extreme longevity of populations—for $\bar{r}K = 1, 5$, and 10 ($\bar{r} = 0.1$/year), the mean extinction time is predicted to be 44, 24,000, and 267 million generations.

Ecologists still do not have a broad understanding of mechanisms of density-dependence for most species. However, the logistic model, in which the expected growth rate of the population gradually declines to zero as the population approaches its carrying capacity, perhaps approximates reality more closely than the exponential model just described, where there is no damping of $\bar{r}$ until K is reached (Begon and Mortimer, 1986). With logistic growth, the mean extinction time is predicted to be $e^{\bar{r}K}(\pi/K)/r \, K)^{1/2}/r$ where $\pi = 3.14$ (Leigh, 1981; Tier and Hanson, 1981; Gabriel and Bürger, 1992). As in the case of exponential growth (previous paragraph), the mean extinction time scales with the product $\bar{r}K$, but at only about half the rate. The reason for this is that under logistic growth conditions, populations are less strongly bounded away from zero density, and as a consequence, recover less rapidly from bottlenecks. For populations with low demographic potential, the mean time to extinction can be substantially less under the logistic growth model than under the exponential growth model. For example, with a carrying capacity of $K = 25$ adults and $\bar{r} = 0.1$/year, the mean extinction times under the logistic and exponential growth models are 140 and 330 generations, respectively.

ganisms), another source of demographic stochasticity can lead to extinction. When the population is small, there is some probability that all of the offspring produced in a generation will be of the same sex. For a population at size N, the probability of this event is $2(0.5^N)$, and the reciprocal of this quantity, 2^{N-1}, gives the mean extinction time if sex-ratio fluctuations are the only source of extinction.[2]

Unless the population is very small, sex-ratio fluctuations alone are unlikely to cause extinction. However, if the population birth rate is a function of the number of females, as is usually the case, sex-ratio fluctuations will generate fluctuations in the population birth rate. This type of synergism can reduce the mean survival time of a population by orders of magnitude relative to expectations from models that ignore sex (Gabriel and Bürger, 1992). For example, if the number of adults the environment can support (K) is less than 25 individuals or so, the mean time to extinction can be as low as 100 generations, even when the maximum rate of population growth is quite high.

The preceding results apply to populations for which the initial density is at the carrying capacity. When a species is recognized as endangered, however, it usually has declined dramatically, at which point the recovery goal is to increase the population density to some higher sustainable level. Richter-Dyn and Goel (1972) developed a general solution for the mean extinction time starting from an arbitrary density, again assuming that random fluctuations in birth and death rates are the only source of extinction risk. Their model is quite flexible in that it allows for any pattern of density-dependence in the birth and death rates.

Random Environmental Changes

Demographic stochasticity becomes less important as the density of a population increases and individual differences average out; however, this is not the case when temporal variation in an exogenous factor, such as the weather, influences the reproductive or survival rates of all individuals in a population simultaneously. Environmental fluctuations influence different individuals to different degrees, but to this point, the theory has only been developed for the situation in which all individuals respond in an identical manner to environmental change. The discussion below expands on the

---

[2]The derivation of this relationship is as follows: the probability that an individual is male is 0.5, and the probability of all individuals being male in a sample of N individuals is 0.5^N. Similarly, the probability that the entire sample consists of females is 0.5^N. Thus, the probability that the sample consists of either all males or all females is $2(0.5^N)$, in which case the population goes extinct due to its inability to reproduce.

preceding section by incorporating environmental as well as demographic stochasticity.

Most models consider the population to be growing with an average growth rate of $\bar{r}$ per capita per year, and variance in this rate among generations, V_e, is due to environmental fluctuations. Typically, it is assumed that the variance is independent of population size and that there is no correlation between the state of the environment in one generation and the next. Such assumptions are probably rarely fulfilled in natural populations, and violations of them would most likely enhance the risk of extinction, as when generations of poor growth conditions tend to be clustered. These caveats aside, a general prediction of models that incorporate environmental stochasticity is that the mean extinction time is determined by the ratio $\bar{r}/V_e$—the higher the average growth rate and the lower the variance, the longer the population is likely to survive. Moreover, the rate of increase of population longevity with increasing K is much slower when environmental stochasticity is present than when demographic stochasticity operates alone (Example 7-2). Depending on the magnitude of V_e relative to $\bar{r}$, even populations with several hundreds or thousands of individuals can be vulnerable to environmental stochasticity.

The theory just discussed treats environmental variation as a factor that drives variation in the intrinsic rate of population growth, $\bar{r}$. Although this is certainly likely to be true in many cases, environmental factors can also define the carrying capacity of a population. Thus, an alternative approach to the treatment of environmental stochasticity is to let K, as well as $\bar{r}$, vary. Variation in K alone cannot cause extinction, unless the carrying capacity actually declines below zero. However, K puts a ceiling on the attainable population size, and bottlenecks in K can magnify the effects of demographic stochasticity by enhancing the variation in the population growth rate due to the smaller sample of reproductive adults. Only limited work has been done on these issues (see Roughgarden, 1975; Slatkin, 1978).

Catastrophes

Catastrophes are extreme forms of environmental variation that suddenly and unpredictably reduce the population size. To the extent that these events are determined by the weather, lightning fires, epidemics, etc., human intervention can do little to influence their frequency. However, because catastrophes affect most members of a population to more or less the same extent, it is clear that, on the basis of chance alone, larger populations will have an increased likelihood of some individuals surviving this kind of event.

Hanson and Tuckwell (1981) and Lande (1993) have considered the time to extinction for populations exposed to randomly occurring events,

---

Example 7-2

In a temporally varying environment, the long-run growth rate, $\tilde{r} = \bar{r} - V_e/2$, governs the vulnerability of a population to extinction (Lewontin and Cohen, 1969; Lande, 1993). Long-run growth rate is the average geometric rate of population expansion. In a constant environment, if a population could expand indefinitely after t time units, its expected size would be $e^{\bar{r}t}$ times its initial size. But in a variable environment, the expected size would be $e^{\tilde{r}t}$ times the initial size. The fact that $\tilde{r}$ can be negative even when the average growth rate $\bar{r}$ is positive underscores the importance of variation in population growth rates—if $(V_e/2) > \bar{r}$, the population will decline towards extinction deterministically. For the case in which the population grows with an expected positive rate $\bar{r}$ as long as it is below the carrying capacity K, the mean time to extinction is proportional to $(1 + KV_e)^{2\bar{r}/Ve}$ (Leigh, 1981). This shows that the scaling of the time to extinction with population size depends on the ratio of the mean to the variance of the rate of population growth, $\bar{r}/V_e$. If this ratio is 1/2, which implies a long-run growth rate of zero, the extinction time is expected to increase linearly with K (as in the case of demographic stochasticity, with $\tilde{r} = 0$). When the variance in the rate of growth is greater than $2\bar{r}$, the extinction time increases less rapidly than linearly with K.

Unless they incorporate all major sources of variability, these models cannot provide reliable estimates of extinction time. To gain some appreciation for the synergistic effects that can arise between demographic and environmental stochasticity, consider the situation in which $K = 100$, $\bar{r} = 0.1$, and $V_e = 0.1$. In this case, the long-run growth rate is slightly positive, $\tilde{r} = 0.05$. In the absence of environmental stochasticity (using results given for demographic stochasticity), the mean extinction time is predicted to be about 27 million generations. On the other hand, a formula derived by Lande (1993), which ignores demographic sources of stochasticity but includes environmental stochasticity, yields a predicted extinction time of 1900 generations. Leigh's formula, which incorporates both sources of variation, predicts a mean extinction time of only 145 generations. Models that further allow for density-dependent (logistic) population growth (Ludwig, 1976; Leigh, 1981; Tier and Hanson, 1981) yield still shorter times to extinction.

---

each reducing the population size to a constant fraction of its current size, the former using a logistic, and the latter an exponential growth model. In these models, there is no demographic or environmental stochasticity of the kinds noted above. Rather, extinction only occurs when, by chance, a cluster of catastrophes occurs. Provided the long-run growth rate is positive, the mean extinction time increases exponentially with the carrying capacity under this model, with the rate of scaling increasing with the frequency of occurrence and magnitude of catastrophes. Assuming catastrophes act locally, spatial subdivision of a species provides a simple means of protection against extinction caused by devastating events.

Accumulation of Deleterious Genetic Factors

The reduction of a population to a low density has several negative genetic consequences that can magnify vulnerability to extinction. Most species harbor far more than enough deleterious recessive genes to kill individuals if they were to become completely homozygous (Simmons and Crow, 1977; Charlesworth and Charlesworth, 1987; Ralls et al., 1988; Hedrick and Miller, 1992). This large genetic load is essentially unavoidable because it is maintained by a deleterious mutation rate of approximately one per individual per generation (Mukai, 1979; Houle et al., 1992). In large populations, deleterious genes, particularly lethal genes, have only minor consequences—the frequencies of most deleterious genes are kept low by natural selection, and their expression is minimal because they are usually masked in the heterozygous state. This situation can change dramatically in small populations. During bottlenecks in population size, mildly deleterious genes, previously kept at low frequency by natural selection, can rise to high frequency by chance. When these genes become completely fixed (reach a frequency of 100%), a permanent reduction in population fitness results.[3]

Although some deleterious genes may be purged from a population early in a population bottleneck (Templeton and Read, 1984), the continued maintenance of a population at small size can only magnify the long-term accumulation of mildly deleterious genes. As noted above, deleterious mutations arise at a rate of about one per individual per generation. Provided the individual selective effects of these genes are small (on the order of $1/4N_e$ or less), they will accumulate at the genomic mutation rate (μ), causing

---

[3]Roughly speaking, if N_e is the effective number of breeding adults and s is the selection intensity opposing a deleterious gene in the homozygous state, then selection is ineffective if $4N_e s < 1$. Typically, because of high variance in family size, the effective population size is a third to a tenth of the actual number of breeding adults (Heywood, 1986; Briscoe et al., 1992). Thus, as a first approximation, if the number of breeding adults is less than $2/s$, natural selection will be essentially incapable of eliminating a deleterious gene—its future frequency will be governed by chance, with the probability of fixation being equal to the initial frequency. The current wisdom is that s for an average mutation is approximately 0.025 (Simmons and Crow, 1977; Houle et al., 1992). Noting that $2/0.025 = 80$, this implies that a substantial number of the rare deleterious genes in a population can drift to high frequency if the number of breeding adults is reduced to 100 or fewer individuals for a prolonged period.

a decline in mean fitness of approximately μs per generation (Lynch, 1994). Thus, if $\mu = 1$ and $s = 0.025$ (as described in footnote 3), a small population would be expected to experience a roughly 2.5% decline in fitness per generation due to deleterious mutations alone, and the rate of mutation accumulation declines with increasing population size. If the effective population size (N_e) is greater than 1,000, mutation accumulation is essentially halted for time scales relevant to endangered species management. However, if the accumulation of deleterious genes reaches the point at which the net reproductive rate of individuals is less than 1, the population is incapable of replacing itself. At this point, the population size begins to decline, and random drift progressively overwhelms natural selection; consequently, decline in fitness accelerates due to the accumulation of deleterious mutations. This synergism, whereby the rate of decline in fitness increases with the accumulation of deleterious genes, has been referred to as a "mutational meltdown" (Lynch and Gabriel, 1990; Lynch et al., 1993) and, once initiated, can lead to rapid extinction.

Loss of Adaptive Variation Within Populations

Most populations, even those undisturbed by human activity, are exposed regularly to temporal and spatial variation in physical and biotic features of the environment. In principle, some species can cope with such selective challenges by simply migrating to suitable habitat (Pease et al., 1989). However, endangered species often live in highly fragmented habitats with inhospitable barriers; migration might not be an option. This leaves adaptive evolutionary change, which requires heritable genetic variation, as the primary means of responding to selective challenges (habitat degradation, global climatic change, species introductions, etc.) that threaten species with extinction.

Consider a population that is faced with a gradual change in a critical environmental factor, such as temperature, humidity, or prey size. If the rate of change is sufficiently slow and the amount of genetic variance for the relevant characters in the population sufficiently high, then the population will be able to evolve slowly in response to the environmental change, without a major reduction in population size. If the rate of environmental change is too high, the selective load (reduced viability and fecundity) on the population will exceed the population's capacity to maintain a positive rate of growth, and although the population might respond evolutionarily, it will become extinct in the process. Thus, for any population, there must be a critical rate of environmental change that allows the population to evolve just fast enough to maintain a stable size. Lynch and Lande (1993) showed that this critical rate is directly proportional to the genetic variance for the character upon which selection is acting.

Several factors influence standing levels of genetic variation for characters associated with morphology, physiology, and behavior. Most forms of natural selection cause a reduction in the genetic variance by eliminating extreme genotypes, the exact amount depending on the intensity of selection. Small populations also lose an expected $1/2N_e$ of their genetic variance each generation due to the chance loss of some genes by random genetic drift. Mutation adds genetic variation to each generation of a population. When populations are kept at a constant size and under constant selective pressures, they ultimately evolve an equilibrium level of genetic variance, at which point the loss due to selection and drift is balanced by mutational input.

For large populations, the magnitude of this equilibrium variation is debatable, because it depends on the gametic mutation rate and the distribution of mutational effects, neither of which are very well understood (Barton and Turelli, 1989). However, for populations with effective sizes of a few hundred or fewer individuals, the expected amount of variation for a typical quantitative character is nearly independent of the strength of selection and proportional to the product of the effective population size and the rate of mutational input of variation (Bürger et al., 1989; Foley, 1992). This implies that for populations containing hundreds or fewer individuals, the rate of environmental change that can be sustained for a prolonged period of time is directly proportional to the effective population size. In other words, a doubling in population size effectively doubles the evolutionary potential of the population.

Some attempts to identify a critical minimum population size for captive populations from a genetic perspective have focused on goals such as the maintenance of 90% of the genetic variation present in the ancestral (predisturbance) population for 200 years (Franklin, 1980; Soulé et al., 1986). Goals of this nature take into consideration the fact that populations that are dwindling in size cannot be in equilibrium. However, these goals are rather arbitrary with respect to choice of acceptable loss and time span. For long-term planning, an alternative approach is to consider that above a certain effective population size, the dynamics of genetic variation are influenced predominantly by selection and mutation, so that any further increase in the effective population size would not significantly influence the amount of genetic variation maintained in the population. Based on the above arguments and because the effective population size is generally severalfold less than the actual number of breeding adults (Heywood, 1986; Briscoe et al., 1992), populations must have about 1,000 individuals to maintain their genetic variation.[4]

---

[4]The actual number depends in part on the biology of the organisms involved, such as sex ratio, breeding behavior, and so on. It can be greater than 1,000 if the effective population size is much smaller than the actual population size.

Habitat Fragmentation

A major area of uncertainty in conservation biology concerns the degree to which population subdivision influences the vulnerability of species to extinction. Even for fairly simple, single-factor investigations in which demographic or environmental sources of randomness are assumed to dominate (Quinn and Hastings, 1987, 1988; Gilpin, 1988), the debate about the effectiveness of a single large reserve as opposed to several small ones is far from being resolved. An advantage of a single large reserve is that it is buffered from demographic stochasticity, but multiple small reserves can buffer an entire species from extinction due to local catastrophes and environmental stochasticity. On the other hand, small isolated populations are precisely the ones that are expected to suffer from inbreeding depression, mutation load, and loss of adaptive potential. Much of the recent theoretical and empirical work on the dynamics of populations with a metapopulation structure can be found in recent volumes by Gilpin and Hanski (1991) and Burgman et al. (1992).

Population subdivision adds another dimension to species viability analysis, because questions are focused not just on the risk of extinction for an individual deme, but for an entire complex of demes. Levins (1970) called a collection of partially or totally isolated populations of the same species a metapopulation, and his early models for site occupancy form the conceptual basis of most current efforts in this area. Levins showed that in an ideal world consisting of an effectively infinite number of subpopulations, each with a constant probability of extinction E and a recolonization rate C, the entire metapopulation will eventually reach an equilibrium with a fraction $1 - E/C$ of the total sites occupied. Because of the randomness of extinction and colonization, the specific sites that are occupied will vary in time.

The intuitive notion behind Levins's work is that unless the extinction rate is zero, the total amount of suitable habitat for a species is unlikely ever to be completely occupied. Elimination of suitable but unoccupied patches of habitat reduces the recolonization rate by making it more difficult for migrants to find suitable sites. Thus, habitat removal could theoretically have the paradoxical effect of increasing the fraction of apparently suitable habitat that is unoccupied, but this is only due to an overall decline in metapopulation size.

Lande (1987) introduced a series of habitat-occupancy models showing that if suitable patches are dispersed to a large enough degree that migrants are unlikely to find them, the local extinction rate will exceed the colonization rate. Thus, there exists a minimum fraction of the total landscape throughout a region that must be suitable for a species to persist. These extinction thresholds, defined by the demographic and dispersal properties of the species, demonstrate that locally abundant species can sometimes be very close to extinction if the proportion of suitable habitat is near the extinction threshold. This again emphasizes that population size alone is not always a good indicator of vulnerability to extinction.

Lande's (1987) models are idealized in that they envision a world consisting of two kinds of habitat patches—hospitable and inhospitable, all of equal size. The real world, of course, is more complex. Patches differ in size and shape, patch quality is usually a continuous variable, and some patches are connected by corridors, others not at all (see Chapter 5). More generalized approaches are discussed by Akçakaya and Ginzburg (1991). A significant feature of their approach is the inclusion of a correlation between the extinction probabilities of adjacent patches. This correlation, if positive, causes a reduction in the expected time to extinction. In other words, if all patches in an area became inhospitable at the same time, there would be no refuges available.

For many species, the adverse consequences of habitat fragmentation are not caused so much by a loss of total area as by changes in the quality of habitat due to the development of edge effects on the margins of reserves (Lovejoy et al., 1986). Edge effects range from microclimatic changes resulting from structural changes in the environment to major alterations in the vegetational community to invasions by exotic species from agricultural and urban settings. The complete impact of edge effects may require several years to develop and may ultimately extend for several kilometers beyond the edge of the reserve. Some attempts have been made to capture the key features of edge effects in mathematical models (Cantrell and Cosner, 1991; 1993). The issues are very complex because they involve interspecific interactions, such as competition between reserve and invading species. Ultimately, the practical application of any of these models requires a deep understanding of the ecology of the species under consideration.

Supplementation

An increasingly common strategy for maintaining wild populations of endangered species is augmentation with stock from breeding facilities, as in the case of hatcheries for Pacific salmonids. An implicit assumption of such procedures is that recipient populations, when they still exist, actually derive some benefit from an artificial boost in population size. There are, however, several reasons why long-term deleterious consequences of supplementation may outweigh the short-term advantage of increased population size.

First, over evolutionary time, successful populations are expected to become morphologically, physiologically, and behaviorally adapted to their local environments. Thus, the introduction of nonnative stock has the potential to disrupt adaptations that are specific to the local habitat. This type of problem takes on added significance when the population employed in stocking has been maintained in captivity. Captive environments are often radically different than those in the wild, and over a period of several generations, "domestication selection" can potentially lead to the evolution of rather different behavioral or morphological phenotypes (Doyle and Hunte, 1981; Frankham and Loebel, 1992; NRC, 1995)—genotypes that perform well in the captive environment are expected to gradually displace those that do not. Furthermore, an overly protective captive breeding program may simply result in a relaxation of natural selection and the gradual accumulation of deleterious genes. For hatchery salmonids, egg-to-smolt survivorship is typically 50% or greater, as compared with 10% or less in natural populations (Waples, 1991; NRC, 1995).

Second, local gene pools can be coadapted intrinsically (Templeton, 1986). Just as the external environment molds the evolution of local adaptations by natural selection, the internal genetic environment of individuals is expected to lead to the evolution of local complexes of genes that interact in a mutually favorable manner. The particular gene combinations that evolve in any local population will be largely fortuitous, depending in the long run on the chance variants that mutation provides for natural selection. The break-up of coadapted gene complexes by hybridization can lead to the production of individuals that have lower fitness than either parental type (outbreeding depression) and takes its extreme form in crosses between true biological species that cannot produce viable progeny. However, outbreeding depression can even occur between populations that appear to be adapted to identical extrinsic environments. The most dramatic evidence comes from reduced fitness in crosses of inbred lines of flies (Templeton et al., 1976) and plants (Parker, 1992), but crosses between outbreeding plants separated by several tens of meters can exhibit reduced fitness (Waser and Price, 1989), as can crosses between fish derived from different sites in the same drainage basin (Leberg, 1993). Outbreeding depression in response to stock transfer is a major concern in the management of Pacific salmon, which are subdivided into demes that are home to specific breeding grounds (Waples, 1991; Hard et al., 1992; NRC, 1995).

Third, augmentation of wild populations with stock from captive breeding programs can have negative ecological or behavioral consequences. Unlike genetic effects, which can take several generations to emerge fully, ecological and behavioral effects can be immediate. For example, high-density hatchery populations of fish are prone to epidemics involving diseases that are uncommon in the natural environment. Such events provide strong selection for disease-resistant varieties of hatchery-reared fish, which subsequently can act as vectors to the wild population. The Norwegian Atlantic salmon is now threatened with extinction resulting from a parasite brought to Atlantic drainages by resistant stock from the Baltic (Johnsen and Jensen, 1986).

Fourth, if a wild population is small because of habitat loss or alteration, the increased population density that results from augmentation can increase competition for food, space, or whatever else the habitat provides. That competition can further reduce the size of the wild population. Harvest of augmented wild populations (particularly if harvest levels are based on total population) can reduce the wild segment of the population unless the harvest effort is directed away from the wild population. A captive breeding and reintroduction program is appropriate only when there is no alternative means of ensuring short-term population viability or when there is strong evidence of historical gene flow. Habitat loss and degradation are the main reasons species become threatened or endangered; therefore, the protection of habitat plays a greater role in preserving these species than captive breeding and reintroduction. For example, as of 1991, the species specialist groups of the International Union for the Conservation of Nature (IUCN), which are international groups of scientists with expertise on specific kinds of animals, had completed conservation plans for 1,370 mammals. Of the recommendations in these plans, 517 concern protecting or managing habitat, while only 19 concern captive breeding and reintroduction (Stuart, 1991).

Captive breeding and reintroduction are appropriate when suitable unoccupied habitat exists and the factors leading to extirpation of the species from this habitat have been identified and reduced or eliminated. Under these circumstances, captive breeding and reintroduction of threatened and endangered species can be part of a comprehensive strategy that also addresses the problems affecting species in the wild (Foose, 1989; Povilitis, 1990; Ballou, 1992; NRC, 1992a). For example, captive breeding and reintroduction enabled the peregrine falcon (*Falco peregrinus*) to repopulate much of North America after the use of DDT was eliminated (Cade, 1990). Similarly, Arabian oryx (*Oryx leucoryx*) were successfully reintroduced in several areas of their original range where hunting was prohibited (Stanley-Price, 1989).

Captive breeding and reintroduction programs should be avoided when possible; however, once the need for a captive breeding program has been identified, it is advisable to initiate it as soon as possible. Starting the program before the wild population has been reduced to a mere handful of individuals increases a program's chances of success. Starting sooner provides time to solve husbandry problems, increases the likelihood that enough wild individuals can be captured to give the new captive population a secure genetic and demographic foundation, and minimizes adverse effects of removing individuals from the wild population.

Captive breeding and reintroduction programs are the most expensive forms of wildlife management (Conway, 1986; Kleiman, 1989) and involve research and management actions. Although genetic and demographic management techniques for captive populations are fairly well developed and can be applied to most species (Ballou, 1992; Ralls and Ballou, 1992), husbandry and reintroduction techniques tend to be species specific. Zoos do not know how to breed many species, such as cheetahs (*Actinomyx jubatus*), reliably in captivity. In such cases, expensive and time-consuming research on genetics, behavior, nutrition, disease, or reproduction might be necessary to find the reasons for lack of breeding success. The reintroduction of captive-bred individuals also poses substantial technical challenges. Considerable research, in captivity and in the field, often is necessary during the early stages of the reintroduction process to develop successful techniques (Kleiman, 1989; Stanley-Price, 1991).

Focusing Conservation Efforts

Life-history models can also help to identify the stages of an organism's life history most likely to be sensitive to conservation efforts. For example, the National Research Council (NRC, 1992b) concluded from life-history data and models that protecting juvenile and sub-adult sea turtles would have a greater effect on increasing population growth than reducing human-caused deaths of eggs and hatchlings. Similarly, by performing an analysis of the sensitivity of the population growth rate of the northern spotted owl to various demographic parameters, Lande (1988), based on the data available then, concluded that the most important contributors to the owl's survival were the adults' annual survival rate, followed by the survival rate of juveniles during their dispersal phase, and annual fecundity.

Distribution of Extinction Times

The preceding discussion summarizes the state of our knowledge of how various factors contribute to the risk of population extinction. For practical reasons, the existing theory focuses almost entirely on the expected time to extinction. However, in the listing and management of endangered species, the primary focus is usually on the likelihood of extinction within a given time frame (Shaffer, 1981, 1987; Mace and Lande, 1991). Risk analysis requires information on the dispersion of the probability distribution of extinction times about the mean. For the models previously cited and many others (Burgman et al., 1992), the distribution of extinction times typically is strongly skewed to the right, with the most likely extinction time (the mode) being substantially less than the mean. In general, it is probably more useful to estimate extinction probabilities as a function of time for different population sizes than to identify some specific MVP.

One conceptually simple way of relating risk to the mean extinction time is to assume that if the current ecological conditions remain stable, the probability of extinction per generation also remains stable.[5] That cannot be strictly true, even in a constant environment, because demographic and genetic sources of stochasticity will ensure that the probability of extinction is not constant in time. For example, if by chance the population size dwindles, the risk of extinction will be elevated above the average risk until the population has recovered to its average size.

LIMITATIONS OF OUR ABILITY TO ESTIMATE RISK

We close this section by again emphasizing that the practical utility of any extinction model depends on the validity of its underlying assumptions. Virtually all work on the vulnerability to extinction has taken a single-factor approach, under the assumption that this will at least yield an understanding of how the expected extinction time scales with population size when a single factor is operating. Other than analytical and computational simplicity, there seems to be little justification for this approach to population viability analysis. Chapter 5 gives some examples of population viability analyses that have been useful and points out the need to recognize the uncertainties discussed here. In nature, populations are exposed to multiple sources of risk simultaneously. Synergism between different risk factors is not reflected in many models, and therefore the risk of extinction can be underestimated, as shown in Example 7-2 (see also Gabriel and Bürger, 1992). A field example of such synergism was described by Woolfenden and Fitzpatrick (1991); epizootic infections of the Florida scrub jay, which reduced local populations by 50%, also lowered reproductive success in the following seasons even after the death rates had returned to normal.

Although analytical results are valuable as guides to research and as methods of comparing the effects of various environmental and management scenarios, they are probabilistic in nature, so they often ignore the

---

[5]In this case, the conditional probability of extinction in any generation (given that the population has survived to that point) is simply the reciprocal of the mean extinction time, i.e., $p_e = 1/\bar{t}_e$ where $\bar{t}_e$ is the mean time to extinction measured in generations. Because the probability that extinction does not occur in $(x - 1)$ consecutive generations is $(1 - p_e)^{x-1}$, and the probability that those $(x - 1)$ generations are immediately followed by extinction is p_e, the probability of extinction in generation x is $p_e(1 - p_e)^{x-1}$. With this approach, the cumulative probability that the population will be extinct by generation t can be computed by solving the preceding expression for $x = 1$ to $x = t$, and summing these probabilities. Results in Gabriel and Bürger (1992) and Tier and Hanson (1981) suggest that this approach might provide a good first-order approximation to the distribution of extinction times due to demographic and environmental stochasticity under a broad range of conditions.

underlying mechanisms. Perhaps their greatest potential is in combination with empirical evidence on extinction times, both in the laboratory and in the field (see for example Pimm et al., 1993). It remains to be seen how relevant such results are to natural populations. Most of the work on vulnerability of species has also focused on nonfragmented populations and, except in the case of asexual populations (Lynch et al., 1993), few formal attempts have been made to incorporate genetics into extinction models. There is a clear need for models that predict distributions of extinction times as a function of population density, demographic rates, mating system, environmental variation, etc. These models, which can only be evaluated by computer simulation (Shaffer and Samson, 1985; Caswell, 1989; Menges, 1992), can be expected to advance substantially in the next few years because computational power is now widely available.

CONCLUSIONS AND RECOMMENDATIONS

• Since the implementation of the Endangered Species Act, numerous models have been developed for estimating the risk of extinction for small populations. Although most of these models have shortcomings, they do provide valuable insights into the potential impacts of various management (or other) activities and of recovery plans. With only a few exceptions, biologically explicit, quantitative models for risk assessment have played only a minor role in decisions associated with the ESA. They should play a more central role, especially as guides to research and as tools for comparing the probable effects of various environmental and management scenarios.

• Despite the major advances that have been made in models for predicting mean extinction times, the existing treatments still have substantial limitations. Most of the models are unifactorial in nature and fail to incorporate the negative synergistic effects that multiple risk factors have on the time to extinction. Efforts to jointly integrate genetic, demographic, and environmental stochasticity into spatially explicit frameworks are badly needed.

• Most extinction models primarily address the mean extinction time. Because decisions associated with endangered species usually are couched in fairly short time frames—less than 100 years—models that predict the cumulative probability of extinction through various time horizons would have greater practical utility.

• Results from population-genetic theory provide the basis for one fairly rigorous conclusion. Small population sizes usually lead to the loss of genetic variation, especially if the populations remain small for long periods. If the members of the population do not mate with each other at random (the case for most natural populations), then the effect of small size on loss of genetic variation is made more severe; the population is said to have a smaller *effective size* than its true size. Populations with long-term mean sizes greater than approximately 1,000 breeding adults can be viewed as genetically secure; any further increase in size would be unlikely to increase the amount of adaptive variation in a population. If the effective population size is substantially smaller than actual population size, this conclusion can translate into a goal for many species for survival of maintaining populations with more than a thousand mature individuals per generation, perhaps several thousand in some cases. An appropriate specific estimate of the number of individuals needed for long-term survival of any particular population must be based on knowledge of the biology of the organisms involved, such as sex ratios, breeding behavior, and so on. If information on the breeding structure of that species is lacking, information about a related species might be useful.

REFERENCES

Akçakaya, H.R., and L.R. Ginzburg. 1991. Ecological risk analysis for single and multiple populations. Pp. 78-87 in Species Conservation: A Population Biological Approach, A. Seitz and V. Loeschcke, eds. Basel, Switzerland: Birkhauser.

Ballou, J.D. 1992. Genetic and demographic considerations in endangered species captive breeding and reintroduction programs. Pp. 262-275 in Wildlife 2001: Populations, D. McCullough and R. Barrett, eds. Barking, U.K.: Elsevier.

Barton, N.H., and M. Turelli. 1989. Evolutionary quantitative genetics: How little do we know? Annu. Rev. Genet. 23:337-370.

Begon, M., and M. Mortimer. 1986. Population Ecology. Sunderland, Mass.: Sinauer Associates.

Briscoe, D.A., J.M. Malpica, A. Robertson, G.J. Smith, R. Frankham, R.G. Banks, and J.S.F. Barker. 1992. Rapid loss of genetic variation in large captive populations of *Drosophila* flies: Implications for the genetic management of captive populations. Conserv. Biol. 6:416-425.

Bürger, R., G.P. Wagner, and F. Stettinger. 1989. How much heritable variation can be maintained in finite populations by a mutation-selection balance? Evolution 43:1748-1766.

Burgman, M.A., S. Ferson, and H.R. Akçakaya. 1992. Risk Assessment in Conservation Biology. New York: Chapman and Hall.

Cade, T. 1990. Peregrine falcon recovery. Endangered Species Update 8:40-45.

Cantrell, R.S., and C. Cosner. 1991. The effects of spatial heterogeneity in population dynamics. J. Math. Biol. 29:315-338.

Cantrell, R.S., and C. Cosner. 1993. Should a park be an island? SIAM J. Appl. Math. 53:219-252.

Caswell, H. 1989. Matrix Population Models. Sunderland, Mass.: Sinauer Associates.

Charlesworth, D., and B. Charlesworth. 1987. Inbreeding depression and its evolutionary consequences. Annu. Rev. Ecol. Syst. 18:237-268.

Conway, W. 1986. The practical difficulties and financial implications of endangered species breeding programs. Int. Zoo Yearbook 24/25:210-219.

Dennis, B., P.L. Munholland, and J.M. Scott. 1991. Estimation of growth and extinction parameters for endangered species. Ecol. Monogr. 61:115-143.

Doyle, R.W., and W. Hunte. 1981. Demography of an estuarine amphipod (*Gammarus lawrencianus*) experimentally selected for high "r": A model of the genetic effects of environmental change. Can. J. Fish. Aquat. Sci. 38:1120-1127.

Foley, P. 1992. Small population genetic variability at loci under stabilizing selection. Evolution 46:763-774.

Foose, T.J. 1989. Species survival plans: the role of captive propagation in conservation strategies. Pp. 210-222 in Conservation Biology and the Black-footed Ferret, U.S. Seal, E.T. Thorne, M.A. Bogan, and S.H. Anderson, eds. New Haven, Conn.: Yale University Press.

Frankham, R., and D.A. Loebel. 1992. Modeling problems in conservation genetics using captive *Drosophila* populations: Rapid genetic adaptation to captivity. Zoo Biol. 11:333-342.

Franklin, I.R. 1980. Evolutionary changes in small populations. Pp. 135-149 in Conservation Biology: an Evolutionary-Ecological Perspective, M. E. Soulé and B. A. Wilcox, eds. Sunderland, Mass.: Sinauer Associates.

Gabriel, W., and R. Bürger. 1992. Survival of small populations under demographic stochasticity. Theor. Pop. Biol. 41:44-71.

Gilpin, M.E. 1988. A comment on Quinn and Hastings: Extinction in subdivided habitats. Conserv. Biol. 2:290-292.

Gilpin, M.E., and I. Hanski, eds. 1991. Metapopulation Dynamics: Empirical and Theoretical Investigations. New York: Academic.

Goodman, D. 1987a. The demography of chance extinction. Pp. 11-43 in Viable Populations for Conservation, M. E. Soulé, ed. New York: Cambridge University Press.

Goodman, D. 1987b. Consideration of stochastic demography in the design and management of biological reserves. Nat. Res. Model. 1:205-234.

Hanson, F.B., and H.C. Tuckwell. 1981. Logistic growth with random density independent disasters. Theor. Pop. Biol. 19:1-18.

Hard, J.J., R.P. Jones, Jr., M.R. Delman, and R.S. Waples. 1992. Pacific Salmon and Artificial Propagation under the Endangered Species Act. NOAA Tech. Memo. NMFS-NWFSC-2. U.S. Department of Commerce, Washington, D.C.

Hedrick, P.W., and P.S. Miller. 1992. Conservation genetics: Techniques and Fundamentals. Ecol. Appl. 2:30-46.

Heywood, J. 1986. The effect of plant size variation on genetic drift in populations of annuals. Am. Naturalist 127:851-861.

Houle, D., D.K. Hoffmaster, S. Assimacopoulos, and B. Charlesworth. 1992. The genomic mutation rate for fitness in *Drosophila*. Nature 359:58-60.

Johnsen, B.O., and A.J. Jensen. 1986. Infestations of Atlantic salmon, *Salmo salar*, by *Gyrodactylus salaris* in Norwegian rivers. J. Fish. Biol. 29:233-241.

Keightley, P.D., and W.G. Hill. 1989. Quantitative genetic variability maintained by mutation-stabilizing selection balance: sampling variation and response to subsequent directional selection. Genet. Res. 54:45-57.

Kleiman, D.G. 1989. Reintroduction of captive animals for conservation. BioScience 39:152-161.

Lande, R. 1987. Extinction thresholds in demographic models of territorial populations. Am. Naturalist 130:624-635.

Lande, R. 1988. Demographic models of the northern spotted owl (*Strix occidentalis caurina*). Oecologia 75:601-607.

Lande, R. 1993. Risks of population extinction from demographic and environmental stochasticity, and random catastrophes. Am. Naturalist 142:911-927.

Lande, R., and G. F. Barrowclough. 1987. Effective population size, genetic variation, and their use in population management. Pp. 87-123 in Viable Populations for Conservation, M.E. Soulé, ed. New York: Cambridge University Press.

Lande, R., and S.H. Orzack. 1988. Extinction dynamics of age-structured populations in a fluctuating environment. Proc. Natl. Acad. Sci. USA 85:7418-7421.

Leberg, P.L. 1993. Strategies for population reintroduction: Effects of genetic variability on population growth and size. Conserv. Biol. 7:194-199.

Leigh, E.G., Jr. 1981. The average lifetime of a population in a varying environment. J. Theor. Biol. 90:213-239.

Levins, R. 1970. Extinction. Pp. 77-107 in Some Mathematical Questions in Biology, M. Gerstenhaber, ed. Providence, R.I.: American Mathematical Society.

Lewontin, R.C., and D. Cohen. 1969. On population growth in a randomly varying environment. Proc. Natl. Acad. Sci. USA 62:1056-1060.

Lovejoy, T.E., R.O. Bierregaard, Jr., A.B. Rylands, J.R. Malcolm, C.E. Quintela, L.H. Harper, K.S. Brown, Jr., A.H. Powell, G.V.N. Powell, H.O.R. Schubart, and M.B. Hays. 1986. Edge and other effects of isolation on Amazon forest fragments. Pp. 257-285 in Conservation Biology: The Science of Scarcity and Diversity, M.E. Soulé, ed. Sunderland, Mass.: Sinauer Associates.

Ludwig, D. 1976. A singular perturbation problem in the theory of population extinction. Soc. Ind. Appl. Math.–Am. Math. Soc. Proc. 10:87-104.

Lynch, M. 1994. Neutral models of phenotypic evolution. Pp. 86-108 in Ecological Genetics, L. Real, ed. Princeton, N.J.: Princeton University Press.

Lynch, M., R. Bürger, D. Butcher, and W. Gabriel. 1993. The mutational meltdown in asexual populations. J. Hered. 84:339-344.

Lynch, M., and W. Gabriel. 1990. Mutation load and the survival of small populations. Evolution 44:1725—1737.

Lynch, M., and R. Lande. 1993. Evolution and extinction in response to environmental change. Pp. 234-250 in Biotic Interactions and Global Change, P. M. Kareiva, J. G. Kingsolver, and R. B. Huey, eds. Sunderland, Mass.: Sinauer Associates.

Mace, G.M., and R. Lande. 1991. Assessing extinction threats: toward a reevaluation of IUCN threatened species categories. Conserv. Biol. 5:148-157.

Menges, E.S. 1992. Stochastic modeling of extinction in plant populations. Pp. 253-276 in Conservation Biology, P.L. Fiedler and S.K. Jain, eds. New York: Chapman and Hall.

Mukai, T. 1979. Polygenic mutations. Pp. 177-196 in Quantitative Genetic Variation, J.N. Thompson, Jr., and J.M. Thoday, eds. New York: Academic.

NRC (National Research Council). 1983. Risk Assessment in the Federal Government: Managing the Process. Washington, D.C.: National Academy Press.

NRC (National Research Council). 1992a. The Scientific Bases for the Preservation of the Hawaiian Crow. Washington, D.C.: National Academy Press.

NRC (National Research Council). 1992b. Decline of the Sea Turtles: Causes and Prevention. Washington, D.C.: National Academy Press.

NRC (National Research Council). 1993. Issues in Risk Assessment. Washington, D.C.: National Academy Press.

NRC (National Research Council). 1995. Upstream: Salmon and Society in the Pacific Northwest. Washington, D.C.: National Academy Press.

Parker, M.A. 1992. Outbreeding depression in a selfing annual. Evolution 46:837-841.

Pease, C.M., R. Lande, and J.J. Bull. 1989. A model of population growth, dispersal and evolution in a changing environment. Ecology 70:1657-1664.

Pimm, S.L., J. Diamond, T.M. Reed, G.J. Russell, and J. Verner. 1993. Times to extinction for small populations of large birds. Proc. Natl. Acad. Sci. USA 90:10871-10875.

Policansky, D. 1993. Application of ecological knowledge to environmental problems: Ecological risk assessment. Pp. 37-51 in Comparative Risk Assessment, C.R. Cothern, ed. Boca Raton, Fla.: Lewis Publishers.

Povilitis, T. 1990. Is captive breeding an appropriate strategy for endangered species conservation? Endangered Species Update 8:20-23.

Quinn, J.F., and A. Hastings. 1987. Extinction in subdivided habitats. Conserv. Biol. 1:198-208.

Quinn, J.F., and A. Hastings. 1988. Extinction in subdivided habitats: Reply to Gilpin. Conserv. Biol. 2:293-296.

Ralls, K., and J.D. Ballou. 1992. Managing genetic diversity in captive breeding and reintroduction programs. Trans. North Am. Wildl. Nat. Resour. Conf. 57:263-282.

Ralls, K., J.D. Ballou, and A. Templeton. 1988. Estimates of lethal equivalents and the cost of inbreeding in mammals. Conserv. Biol. 2:185-193.

Richter-Dyn, N., and N.S. Goel. 1972. On the extinction of a colonizing species. Theor. Pop. Biol. 3:406-433.

Risk Assessment Forum. 1992. Framework for Ecological Risk Assessment. EPA/630/R-92/001. U.S. Environmental Protection Agency, Washington, DC.

Roughgarden, J. 1975. A simple model for population dynamics in stochastic environments. Am. Naturalist 109:713-736.

Shaffer, M.L. 1981. Minimum population sizes for species conservation. BioScience 31:131-134.

Shaffer, M.L. 1987. Minimum viable populations: Coping with uncertainty. Pp. 69-86 in Viable Populations for Conservation, M.E. Soulé, ed. New York: Cambridge University Press.

Shaffer, M.L., and F.B. Samson. 1985. Population size and extinction: A note on determining critical population sizes. Am. Naturalist 125:144-152.

Simmons, M.J., and J.F. Crow. 1977. Mutations affecting fitness in Drosophila populations. Annu. Rev. Genet. 11:49-78.

Slatkin, M. 1978. The dynamics of a population in a Markovian environment. Ecology 59:249-256.

Soulé, M.E., M. Gilpin, N. Conway, and T. Foose. 1986. The millenium ark: How long a voyage, how many staterooms, how many passengers? Zoo Biol. 5:101-114.

Stanley-Price, M.R. 1989. Animal Reintroductions: The Arabian Oryx in Oman. Cambridge, U.K.: Cambridge University Press. 291 pages.

Stanley-Price, M.R. 1991. A review of mammal re-introductions, and the role of the reintroduction specialist group of IUCN/SSC. Pp. 9-25 in Beyond Captive Breeding: Reintroducing Endangered Mammals to the Wild, J.H.W. Gipps, ed. Zoological Society of London Symposia 62. Oxford, U.K.: Clarendon Press.

Stuart, S.N. 1991. Re-introductions: To what extent are they needed? Pp. 27-37 in Beyond Captive Breeding: Reintroducing Endangered Mammals to the Wild, J.H.W. Gipps ed. Zoological Society of London Symposia 62. Oxford, U.K.: Clarendon Press.

Templeton, A.R. 1986. Coadaptation and outbreeding depression. Pp. 105-116 in Conservation Biology: The Science of Scarcity and Diversity, M.E. Soulé, ed. Sunderland, Mass.: Sinauer Associates.

Templeton, A.R., and B. Read. 1984. Factors eliminating inbreeding depression in a captive herd of Speke's gazelle. Zoo Biol. 3:177-199.

Templeton, A.R., C.F. Sing, and B. Brokaw. 1976. The unit of selection in Drosophila mercatorum. I. The interaction of selection and meiosis in parthenogenetic strains. Genetics 82:349-376.

Thompson, G.G. 1991. Determining Minimum Viable Populations under the Endangered Species Act. NOAA Tech. Memo. NMFS F/NWC-198. U.S. Department of Commerce, Washington, D.C.

Tier, C., and F.B. Hanson. 1981. Persistence in density dependent stochastic populations. Math. Biosci. 53:89-117.

Tuljapurkar, S. 1989. An uncertain life: Demography in random environments. Theor. Pop. Biol. 35:227-294.

Waples, R.S. 1991. Genetic interactions between hatchery and wild salmonids: Lessons from the Pacific Northwest. Can. J. Fish. Aquat. Sci. 48:124-133.

Waser, N.M., and M.V. Price. 1989. Optimal outcrossing in *Ipomopsis aggregata*: seed set and offspring fitness. Evolution 43:1097-1109.

Woolfenden, G.E., and J. Fitzpatrick. 1991. Florida Scrub Jay Ecology and Conservation. Pp. 542-565 in Bird Population Studies: Relevance to Conservation and Management, C.M. Perrins, J.D. Leberton, and G.J.M. Hirons, eds. New York: Oxford University Press.

Wright, S. 1931. Evolution in mendelian populations. Genetics 16:97-159.

Zeng, Z.-B., and C.C. Cockerham. 1991. Variance of neutral genetic variances within and between populations for a quantitative character. Genetics 129:535-553.

8

Making ESA Decisions in the Face of Uncertainty

The previous chapter described our ability to estimate the risk of extinction for populations of organisms. This is a major part of estimating the degree of endangerment of a population or species. In this chapter, we build on that information to discuss how important understanding and assessing risk is in ESA decision making and the question of whether different levels of risk should apply to different decisions. Finally, because decisions regarding endangered species must always be made in the face of uncertainty regarding estimates of extinction risk and future events, we suggest ways of improving agency decisions involving risk and uncertainty.

DECISIONS REQUIRED UNDER THE ESA

The objectives of the ESA are to conserve the ecosystems upon which endangered and threatened species depend, provide a program for the conservation of endangered and threatened species, and achieve the purposes of several international conservation agreements. While these objectives are not intrinsically or philosophically conflicting, they can conflict when agencies faced with limited budgets must decide how to apportion funds. More serious conflicts arise when the objectives of the act conflict with other human objectives, such as private-property rights and private and public developments. The act specifies the extent to which such conflicting objectives should be considered when making the different decisions required under the act. Consideration of human objectives other than those specified in the act is specifically prohibited when making decisions regarding listing, take, and jeopardy, but is required when making decisions regarding critical habitat. Initial recovery planning is to be based solely on scientific considerations, but economic effects of the plan are to be considered before implementation.

The terminology of the act implies that many decisions regarding conservation of species should consider estimates of extinction risk. Specific examples of such terminology include the definitions of endangered and threatened species, the provisions for removing species from the list, and the definitions of jeopardy on public lands and taking on private lands (see Box 8-1).

THE NEED FOR NEW APPROACHES TO DECISION MAKING

Agency decisions that have been taken under the direction of the act have been criticized by the general public (Mann and Plummer, 1992) and

---

Box 8-1
Examples of terminology in the ESA (italicized) that implies an assessment of the degree of risk to a species

ENDANGERED SPECIES: any species which is *in danger of extinction* throughout all or a significant portion of its range.

THREATENED SPECIES: any species which is *likely to become an endangered species* within the foreseeable future throughout all or a significant portion of its range.

CRITICAL HABITAT: the specific areas within the geographical area occupied by the species . . . on which are found those physical or biological features *essential to the conservation of the species* and which may require special management considerations or protection.

REMOVAL FROM LIST (RECOVERY): [each recovery plan shall include] objective, measurable criteria which, when met, would result in a *determination*, in accordance with the provisions of this section, *that the species be removed* from the list.

JEOPARDY ON PUBLIC LANDS: [each federal agency shall ensure that any action authorized, funded or carried out by such agency] is not *likely to jeopardize the continued existence* of any endangered species or threatened species or result in the destruction or adverse modification of habitat of such species which is determined to be critical.

TAKE ON PRIVATE LANDS: . . . taking will not *appreciably reduce the likelihood of the survival and recovery* of the species in the wild.

the scientific community (Brownell et al., 1989; Goodman, 1993; Wilcove et al., 1993, Tear et al., 1993) for being arbitrary. For example, Brownell et al. (1989) pointed out that, at least for cetaceans, the list of threatened and endangered species is not scientifically defensible. Several large whales in little danger of extinction were on the list in 1989, although many small dolphins and porpoises with very small population sizes were not. Population size alone is not necessarily a sensitive indicator of extinction risk, but many species had declined to alarmingly low numbers by the time they were listed. Wilcove et al. (1993) found that the median population sizes of species when listed were about 1,000 individuals for animals and 120 individuals for plants.

The backlog of candidate species awaiting listing and the number of legal actions taken related to listing decisions indicate that the current process for making listing decisions needs review and revision. The Fish and Wildlife Service (FWS) uses three categories to help make such decisions (see Box 8-2). Category 1 now contains about 400 species for which the agencies have substantial information to support the proposal to list but do not have sufficient resources to complete the process. Category 2 now contains about 3,500 species for which a petition to list might be justified but is deemed to be lacking in critical information. The large number of species in Category 2 indicates that the agencies have not developed a workable process for making listing decisions when faced with limited data, uncertainty, and limited financial and human resources. Indeed, lack of scientific information and uncertainty plague many public and private policy decisions.

Recovery plans often fail to provide appropriate guidance on biologically reasonable levels of risk. A recent survey of recovery plans concluded that goals for species recovery were often unrealistically low and that these plans frequently manage for extinction rather than survival (Tear et al., 1993). For example, of the 54 threatened and endangered species for which population size data were available, 15 (28%) had recovery goals set at or below the existing population size at the time the plan was written (Tear et al., 1993). If the population was endangered at the time of listing—and the review of Wilcove et al. (1993) suggests that most listed populations are at risk—then such recovery goals are probably too low.

The committee also notes that some recovered species have not been delisted or upgraded (i.e., from endangered to threatened) in a timely fashion. For example, eastern Pacific gray whales (*Eschrictius robustus*) were not delisted until 1994 (*Fed. Reg.* 59:31094), although population estimates were approaching the likely pre-exploitation sizes of 12,000-15,000 by the late 1970s and have continued to increase (Reilly, 1992). The comments about decisions to list also apply to decisions about delisting.

Box 8-2
U.S. Fish and Wildlife Service definitions for Categories 1, 2, and 3, under the Endangered Species Act (*Fed. Reg.* 58:51145, Sept. 30, 1993).

Category 1.—Taxa for which the Service has on file sufficient information on biological vulnerability and threat(s) to support proposals to list them as endangered or threatened species. Proposed rules have not yet been issued because this action is precluded at present by other listing activity. Development and publication of proposed rules on Category 1 taxa are anticipated, however, and the Service encourages other Federal agencies to give consideration to such taxa in environmental planning.

Category 2.—Taxa for which information now in the possession of the Service indicates that proposing to list as endangered or threatened is possibly appropriate, but for which sufficient data on biological vulnerability and threat are not currently available to support proposed rules. The Service emphasizes that these taxa are not being proposed for listing by this notice, and that there are not current plans for such proposals unless additional supporting information becomes available. Further biological research and field study usually will be necessary to ascertain the status of taxa in this category. It is likely that many will be found not to warrant listing, either because they are not threatened or endangered or because they do not qualify as species under the definitions in the Act, while others will be found to be in greater danger of extinction than some taxa in Category 1.

Category 3.—Taxa in Category 3 are not current candidates for listing. Such taxa are further divided into three subcategories to indicate the reason(s) for their removal from consideration:

Category 3A.—Taxa for which the Service has persuasive evidence of extinction. If rediscovered, such taxa might acquire high priority for listing. At this time, however, the best available information indicates that the taxa in this subcategory, or the habitats from which they were known, have been lost.

Category 3B.—Names that, on the basis of current taxonomic understanding (usually as represented in published revisions and monographs), do not represent distinct taxa meeting the Act's definition of "species." Such supposed taxa could be reevaluated in the future on the basis of new information.

Category 3C.—Taxa that have proven to be more abundant or widespread than previously believed and/or those that are not subject to any identifiable threat. If further research or changes in habitat indicate a significant decline in any of these taxa, they may be reevaluated for possible inclusion in categories 1 or 2.

PROVIDING OBJECTIVE RISK STANDARDS

The qualitative definitions in the act provide a framework for decision-making, that is, they provide a list of administrative and management actions and a general rationale for selecting each action. However, qualitative definitions alone can be interpreted in different ways by different people, and agencies have provided no guidance on the appropriate degrees of extinction risk for making the different decisions required by the act, such as listing a species as either threatened or endangered or declaring a species recovered. To ensure that ESA decisions protect endangered species as they are intended to and do so in a scientifically defensible way requires objective methods for assessing risk of extinction (see Chapter 7) and for assigning species to categories of protection according to their risk of extinction. Standards for assigning species to categories should be quantitative wherever possible, and, when this is not possible, qualitative procedures should at least be systematic and clearly defined.

Developing Quantitative Risk Standards

Agencies could achieve greater consistency in decisions if they provided quantitative risk standards for such terms as "endangered" and "threatened" to the many agency personnel involved in implementing the act. Risk levels should be defined as the probability of extinction within a specific time. As an example of quantitative guidance, Shaffer (1981) defined a minimum viable population size as that which would have at least a 99% chance of surviving for 1,000 years. The committee is not endorsing Shaffer's definition but noting that by providing quantitative guidance, Shaffer has made it possible to discuss or disagree with his definition objectively and apply the definition in a standard manner. As another example of a quantitative definition, Mace and Lande (1991) suggested that endangered could be defined as a 20% chance of extinction in 20 years.

Time Frame for Estimating Risk of Extinction

When providing a quantitative standard for assigning risk categories, risk of extinction must be defined with a specific time frame in mind, i.e., x probability of extinction in y years. Critical levels of extinction risk, which will trigger ESA decisions, such as listing or delisting, must be associated with a particular number of years or generations. As the time frame increases, the probability of extinction also increases, approaching 100% for all species if the period is long enough (see Chapter 2). The choice of a time frame for evaluating risk of extinction for purposes of the ESA reflects scientific and societal concerns. From a scientific standpoint, time should be long enough to guard against making management choices that might have favorable effects over the short term (e.g., the next 5 to 10 years), but consequent adverse effects over the longer term (e.g., the next 100 years or more). An example of such a choice might be management that extends the lifespan of adults currently in a population (enhancing short-term survival), but jeopardizes successful reproduction (leading to population decline as the current adult population ages and dies). Species' life-history characteristics help determine an appropriate time frame, with longer times being more appropriate for longer-lived species.

Another scientific consideration is the time scale for natural cycles of disturbance and regeneration in the species' habitat. Evaluating the success of endangered species management over only a portion of the natural habitat cycle runs the risk of confusing natural fluctuations in population size with adverse reactions to management.

Societal considerations regarding time frame include the desire to preserve species and their habitats over time scales meaningful to humans and their offspring, political cycles of 2-6 years, economic cycles, and many others. Because of the large variety of societal factors, this committee cannot specify all the appropriate time scales that should be considered in decision making or even the range of time scales for which extinction probabilities should be calculated. However, it is clear that the time scales implicit in the various calculations should be made explicit for informed decision making. In some cases, where there is an immediate and potentially reversible threat to species survival (such as a proposed development), it could make sense to analyze the probability of survival over a short period, perhaps 5 to 10 years or less, when comparing options for species recovery over the short term. Such an analysis should be followed, however, by another assessment of extinction risk over a longer period to ensure that short-term gains do not become long-term losses.

Listing Systems Based on Objective Criteria and Rules

A system for making listing decisions requires a set of objective criteria for assigning species to risk categories, such as endangered or threatened. The objective criteria most suitable for making listing decisions would be different degrees of extinction risk. However, we rarely have sufficient data to allow good estimates of extinction risk, so we need a system allowing the use of other criteria as well. Such a system could be based on objective criteria, such as some combination of population size and number, believed to represent a specific level of extinction risk.

The need to develop a listing system based on objective criteria has been recognized by the principal international organization concerned with the listing and conservation of endangered species, the International Union for the Conservation of Nature (IUCN). In June 1992, the CITES (Convention on the International Trade in Endangered Species of Flora and Fauna) Standing Committee requested that IUCN help to develop new criteria for listing species in the appendices to the CITES treaty, which regulates trade in wildlife and wildlife products. The resulting proposed IUCN system for assessing degree of threat (Mace et al., 1992) was produced after an international effort by groups of scientists in a series of workshops. The proposed system is now being evaluated by members of the IUCN/SSC taxon specialist groups, various CITES committees, and interested scientists (Gnam, 1993) and was considered at the CITES 1994 meeting. At that meeting, a proposal was made to adopt the IUCN criteria; a counterproposal was made by the United States and a working group developed a compromise, which was recently accepted by CITES (Rosemarie Gnam, FWS, pers. commun., March 1, 1995). Much of the substance of the IUCN criteria remains in Annex 5 of the report as definitions, rather than the criteria for listing.

In the proposed IUCN system, a species could be listed as endangered based on any of several criteria, each of which was intended to represent approximately the same risk of extinction. Decisions to list a species could be based on any of the following criteria: probability of extinction, trends in abundance, population size, number of populations, and geographical extent (Mace et al., 1992). The IUCN system was based on a combination of two previous approaches for assessing degree of threat: population-based criteria developed by those working with higher vertebrates and area-habitat criteria developed by those working with plants and invertebrates. Population-based criteria, known as the "Mace-Lande criteria," were proposed in 1991 (Mace and Lande, 1991). They suggested three categories:

Critical: 50% probability of extinction within 5 years or two generations, whichever is longer.

Endangered: 20% probability of extinction within 20 years or 10 generations, whichever is longer.

Vulnerable: 10% probability of extinction within 100 years.

Unfortunately, due to the limited data on most taxa of conservation concern, a formal population viability analysis to estimate the probability of extinction in a particular case is often impossible. Moreover, population viability models for plants are poorly developed (Menges, 1990). To address this problem, Mace and Lande offered several surrogate criteria based on other types of information. For example, they considered that a population would qualify as endangered if the current population estimate was fewer than 250 mature individuals, even if insufficient demographic data were available to calculate an estimated probability of extinction. The current IUCN criteria rely heavily on population viability analysis, population size, and range area. Concern has been expressed that these criteria omit historical data on a species' abundance and distribution, omit data on reproductive fitness, and ignore life cycles (Gnam, 1993). Furthermore, the IUCN system is still incomplete, because it lacks a set of rules to allow decisions to be made when there is uncertainty regarding the criteria used to categorize species according to relative risk of extinction. Nevertheless, those criteria represent the most important scientific effort to date to reach consensus on standard criteria for assigning taxa to threat categories in a uniform, objective manner.

The adoption of a similar system in the U.S. would make listing decisions more consistent. Unfortunately, there is as yet no evidence that the criteria used in the IUCN system do represent comparable degrees of risk. Any system of criteria developed for use with the ESA should be thoroughly tested against a variety of population and metapopulation models before implementation. Designing and testing an appropriate system for listing species is a formidable scientific task best accomplished by an independent scientific committee. For example, for a much simpler problem, management of the commercial harvest of whales, the Scientific Committee, of the International Whaling Commission took 6 years to evaluate five alternative proposed management procedures (Donovan, 1989; R. Brownell, pers. commun., NMFS, 1994).

Limitations on Estimates of Risk

Our ability to estimate risk of extinction for use in assigning species to protection categories is limited by our understanding of the factors influencing extinction. Two areas where we are acutely aware of limitations are in models used to estimate risk of extinction and in our understanding of the role of critical thresholds of risk and of cumulative effects of risk factors.

Limits of Models

As described in Chapter 7, models for estimating risk of extinction are limited in their ability to incorporate the full complexity of species population dynamics. Estimates of risk derived from these models may reflect only a subset of the factors actually influencing a species' risk of extinction. As discussed in Chapter 7, it seems likely that the simplifications and omissions of current models can underestimate risk of extinction.

Poor Understanding of Cumulative Effects and Thresholds

Decisions under the ESA regarding take or jeopardy require making decisions regarding incremental risks of extinction. Assessing the added risk from specific human actions is usually an even more difficult task than estimating the overall extinction risk to a species. Individual human actions, such as developing a few acres of habitat, pose small incremental risks of extinction. At some point, however, the small incremental risks from numerous human actions, if not stemmed, will accumulate so as to produce a major effect. Not enough is known about cumulative effects and threshold points to make accurate risk predictions possible (e.g., Beanlands et al., 1986), although there has been some theoretical work on critical thresholds of habitat patch size or fragmentation (e.g., Lande, 1988 for spotted owl habitat fragmentation). When considering the probable effects of incremental human activities, it is reasonable to assume that additional activity means additional risk, but we rarely know whether the relationship between additional activity and additional risk is linear or whether there might be critical levels of activity above which the risk of extinction increases dramatically.

Should Different Risk Standards Apply to Different ESA Decisions?

Endangered, Threatened, Recovered

In judging whether different levels of risk should apply to different types of decisions under the act, the committee considered carefully the terminology of the act shown in Box 8-1. The definition of an endangered species as one that is already in danger of extinction and a threatened species as one that is likely to become an endangered species implies that a species listed as endangered is at greater risk of extinction than a threatened one. The determination that a species should be removed from the list implies that its risk of extinction has decreased to the point where it is no longer considered threatened. Thus, it is clear that determinations of a single species as successively endangered, threatened, and recovered should represent a series of decreasing levels of risk of extinction faced by the species.

Different Taxa

Although cross-species comparisons are complicated by many factors specific to the biology of individual species, it is appropriate to set the same degree of risk as a standard for listing any species, whether plant or animal, as endangered and another, somewhat lower degree of risk, for listing any species as threatened. Thus it is reasonable to expect that species determined to be endangered should, on the average, face a greater risk of extinction than those determined to be threatened. This seemingly obvious ordering of risk has not always been followed in practice (Wilcove et al., 1993).

Public Versus Private Land

Controversy has arisen over whether the inclusion of habitat destruction or modification as a form of taking under Section 9 sets a different standard of responsibility for protection of endangered species by private versus public entities. (We note again that this interpretation of taking is under court review at this writing.) In particular, it has appeared to some that the standard of responsibility might be interpreted to be more stringent for private than for public entities. This seems, if anything, the reverse of what was intended by the ESA. From a scientific perspective, actions resulting in a given degree of risk of extinction are equally hazardous to species whether they are carried out by public or private entities on public or on private land. The committee sees no scientific reason for setting different standards for categorizing risk of extinction under different sections of the act. However, because public and private entities behave differently, achieving the same degree of biological protection on public versus private lands does not necessarily imply identical regulatory requirements on behalf of species experiencing the same risk of extinction on public and on private lands.

USING STRUCTURED APPROACHES TO DECISION MAKING

Why use structured approaches to ESA decision making? Because the issues are complex and relevant scientific data are often fragmentary, it can be difficult to make decisions regarding endangered or threatened species. Decisions regarding endangered species are often characterized by insufficient data, probabilistic predictions regarding future events, considerable uncertainty regarding the accuracy of these predictions, conflicting management objectives, disagreement over the best course of action, and the need to justify whatever decision is made.

Given these problems, it is important that we try to make our decision-making process as explicit as possible, especially because research on the psychology of human decision making (Hogarth, 1980; Kahneman et al., 1982; von Winterfeldt and Edwards, 1986) has shown that intuitive decisions exhibit many inconsistencies and biases, particularly when probabilistic information and multiple objectives are involved. The use of an explicit framework to guide decisions can help us make choices that are consistent with our goals, data, and beliefs and facilitate compromise among those with differing views. Several techniques developed in the fields of operations research and management science—the academic disciplines dealing with scientific approaches to decision making—provide helpful frameworks for making these difficult decisions. These techniques help decision makers think about the decision in a systematic way; break down difficult decisions into a series of smaller, easier decisions; and document the process used to reach the decision, which makes the decision easier to justify and defend.

Some applications of these structured approaches, particularly the qualitative ones, might appear to be simple pleas for clear thinking, and they are. But clear thinking does not come easily or naturally in the face of scientific complexity and uncertainty, competing objectives within recovery programs and beyond, and political pressures from multiple constituencies. We propose decision analysis as an example of a structured problem-solving method, although other methods (such as the Analytic Hierarchy Process (Saaty, 1990)) or approaches based on goal programming and multiattribute ranking (Ralls and Starfield, 1995) can be useful as well. We stress the merits of using these tools as conceptual frameworks, not just as number-crunching devices. And we emphasize that using these approaches is not necessarily a call for more information, but rather for more coherent use of existing information.

Using Subjective Probability and Expert Opinion

In some cases, there is very little "hard" information that seems relevant to estimating the risks affecting endangered species, but some experts have accumulated experience that allows them to make informed judgments about these risks. Such expert judgment is so often available for endangered species decisions that it is of great benefit to have orderly methods of eliciting and using it for decision making. One of the strengths of decision analysis is its ability to estimate "subjective probabilities" and then use them for analysis in the same way that long-term frequency estimates would be used (Behn and Vaupel, 1982; von Winterfeldt and Edwards, 1986; Maguire, 1987). The methods allow a concrete expression of expert opinion, facilitating scrutiny by the public and comparison with other views.

Another place where expert opinion can be essential is in cases when some background information is available, but unique aspects of a situation differentiate it from similar situations that have occurred. An example might be assessing the likelihood of successful reproduction in a captive population where experience has been dismal but new reproductive technologies have been developed that might prove helpful, as for example with the endangered Hawaiian crow (NRC, 1992a). In this case, it is relevant to amend or update the historical information in light of the new techniques. There are formal ways to combine old data with new opinion (and new data with old opinion) using Bayesian estimation (Raiffa, 1968; Clemen, 1991).

Linking Science and Values

Even though estimates of risk are grounded in scientific information, those implementing the act often make value judgments when making decisions about listing, jeopardy, etc. They are deciding which species to list quickly and which to relegate to delayed listing, which areas of habitat are worth the socioeconomic cost or political effort to designate as critical and which are not, what degree of jeopardy is worth altering a federally funded development project for and what is not. Many citizens are willing to allow public officials to make such judgments on their behalf, but those involved might be more comfortable if the values informing those judgments and their effects on ESA decisions were articulated more clearly.

A hallmark of formal decision analysis (Behn and Vaupel, 1982; Clemen, 1991) and other structured problem-solving methods (Ralls and Starfield, 1995) is an emphasis on articulating clearly the objectives for a decision and criteria for evaluating how well alternative proposals might meet those objectives. Use of such methods can improve ESA decisions by making the connection between values, objectives, and decisions more transparent, helping to disarm criticisms that the government is capricious or partisan in implementing the act (Mann and Plummer, 1992; Tear et al., 1993).

Making good use of science, as instructed in the ESA, requires making appropriate connections between the values and objectives being pursued in a decision and the scientific evidence and reasoning used to evaluate alternative ways of meeting those objectives. Science by itself is not sufficient input to policy decisions, apart from the objectives and values it serves. Articulating an explicit framework can help link science and values and lead to better and more defensible decisions.

Scientific Uncertainty in ESA Decisions

For even the best-studied endangered species, essential pieces of information might be lacking, yet decisions must be made. Sometimes, it is possible to delay action while gathering better information, although that strategy carries its own risks. Sometimes, important factors affecting how management actions turn out, such as catastrophic weather conditions or pollution accidents, are inherently uncertain, and no amount of further study could do more than improve the accuracy and precision of estimates of their likelihoods. In any case, weighing the best choice under uncertainty about outcomes is a necessity. This kind of probabilistic reasoning does not come naturally and many managers are uncomfortable with it, resorting to short-cut heuristics to simplify information and justify their choices (von Winterfeldt and Edwards, 1986). The framework of decision analysis offers a structure for considering probabilistic information in a coherent and consistent way, providing better use of whatever information is available to guide decisions.

Estimating Uncertain Quantities

Several types of information bearing on ESA decisions can be uncertain, including factors influencing how management actions turn out (such as weather events or sociopolitical events) and measures of outcome (such as the likelihood of population persistence under a particular management strategy). Such probabilistic quantities enter the analysis in different ways, but in both cases, methods of estimating those quantities are needed. Sometimes, relevant long-term data on the frequency of an uncertain event can be used to estimate its probability; as an example, there are weather records for severe storms in particular coastal areas. These could be used to estimate the likelihood of a tropical storm striking the Gulf Coast of Texas near the wintering ground for the migratory whooping crane population or the probability that a hurricane would destroy the habitat of some red-cockaded woodpeckers. In one case, population models were used to help decisions concerning conservation of endangered sea turtles by identifying the most sensitive life stages (NRC, 1992b).

In other cases, no useful long-term data are available, but there might be models incorporating our best understanding of the factors affecting the likelihood of an uncertain event. Examples are stochastic population models, such as are discussed in Chapter 7. Those use information about demographic parameters and environmental factors to predict probability of extinction for a population with particular characteristics.

Reducing Uncertainty by Gathering Information

Listing actions, recovery plans, or other ESA decisions often are delayed due to inadequate information. Those implementing the act almost always believe that with additional information, they could make a better decision. Nevertheless, decisions to delay action pending further information and directives to gather additional information should be viewed critically. What kind of information and of what quality could be gathered within the time and resources available? What are the possible answers that such investigation might reveal? What decisions would be triggered by different answers? How are those decisions different from those that would be made using existing information? What effect will continuing the status quo have on species status and on options for future action? Considering these questions in a structured framework can make it more likely that a reasonable decision will be made.

An example of such an analysis comes from Bart and Robson (in press).

They analyzed the variability in raptor population data to find out how many years of data would be required to tell whether the northern spotted owl population was stable or increasing, a criterion for delisting specified in the spotted owl recovery plan. Their analysis showed that it would not be possible to render a sound judgment about delisting based on fewer than 8 years of population monitoring data. (See Box 8-3 on statistical power and types of errors.) Such an analysis promotes realistic expectations about the time and effort required to obtain a satisfactory answer and forestalls charges that the FWS is delaying delisting for nonscientific reasons.

Sometimes a qualitative but orderly consideration of questions is sufficient to guide action, giving managers the confidence either to pursue additional information or proceed on the basis of the information they already have. At other times, a more formal analysis of the value of information (Raiffa, 1968; Clemen, 1991) might be needed. In either case, the scientific uncertainty must be examined within the context set by the objectives for a particular situation. The question of how many years of monitoring data would be required to make a decision about delisting the northern spotted owl can be answered only with reference to the level of confidence required, which can be determined only with reference to the objectives of spotted owl management (Taylor and Gerrodette, 1993).

More information almost always seems better to those trained as cautious natural resource scientists. Yet, too much risk aversion, or fear of making the wrong decision based on limited information, can be crippling. The California condor and the black-footed ferret are good examples. In these cases, and in many similar ones worldwide, a management decision to remove individuals from the wild to begin or to augment a captive-breeding effort would have an indisputably negative effect on the survival of the population in the wild (Maguire, 1986; 1989). Such a decision must be justified in terms of the likelihood of extinction of the population even if no removals were made, and the long-term benefit to species survival from a captive breeding program, if successful. In the condor and ferret programs, only when it became clear that a continuation of status-quo management of the wild population would lead to disaster did the negative consequences of removals become acceptable. But by that point, it was very nearly too late to make a success of the captive-breeding programs (Ralls and Ballou, 1992). Although both programs were ultimately successful in producing animals for reintroduction, neither case can be cited as a model of informed decision making under uncertainty. In both cases, too much attention was given to the possible negative consequences of the novel strategy (capturing animals from the wild) and too little to the possible negative consequences of continuing the conservative approach (trying to protect the dwindling wild populations).

Box 8-3
Statistical power and types of errors

Statistical tests are used to control the likelihood of drawing erroneous conclusions about scientific hypotheses from data. The null hypothesis usually states that there is no difference between two (or more) populations with regard to some characteristic in which we are interested. In contrast, the alternative hypothesis states that there is a difference in that particular characteristic.

The alternative hypothesis is the more important one from the practical point of view because we would be likely to take some management action if it were true. However, the rules of statistics require that we test the null hypothesis rather than the alternative hypothesis. In other words, we make an "argument by contradiction:" we assume that the null hypothesis is true and then see if the data enable us to conclude that it is actually highly likely to be false. If so, we conclude that the alternative hypothesis is likely to be true. On the other hand, if we cannot show that the null hypothesis is likely to be false, we conclude that the alternative hypothesis is unlikely to be true.

Because statistical tests merely allow us to estimate the probability that a hypothesis is true given a particular set of observations, two types of mistakes are possible. The first possible mistake is that we might conclude that the null hypothesis is false when in fact it is true. This is called rejecting a true null hypothesis or making a Type 1 error. The second possible mistake is that we might conclude that the null hypothesis is true when in fact it is false. This is called accepting a false null hypothesis or making a Type 2 error.

A specific example may be helpful. Suppose we would like to know whether or not the population of an endangered species is smaller in 1990 than it was in 1980. There would be no need for a statistical analysis if we could count every individual: if the number counted in 1990 was smaller than the number counted in 1980 we could conclude with certainty that the population had decreased over this time period. The need for statistics arises because we cannot count every individual; we can only conduct surveys and make estimates of population size. These estimates might be very precise, that is we might have a low degree of uncertainty about them. Usually, however, we are only able to make somewhat imprecise estimates about which we have a considerable degree of uncertainty. Regardless of the degree of precision of our estimates, we must consider a probability distribution of possible population sizes in 1990 and another probability distribution of population sizes in 1980 (Figure 8-1a for a more precise estimate and Figure 8-1b for a less precise estimate).

The null hypothesis is that there is no difference between the population size in 1980 and that in 1990. The alternative hypothesis is that the population size in 1990 is smaller than the population size in 1980. We test the null hypothesis that there is no difference between the population sizes in 1980 and in 1990.

Three things determine our ability to say whether the 1990 population is smaller than the 1980 population: the magnitude of the difference between our two population estimates; the precision of our population estimates; and the risk we are willing to take that we falsely conclude the 1990 population is smaller. The situation with very precise population estimates is shown in

Box 8-3 Continued

Figure 8-1a. Looking at the distribution of population estimates for 1980, we see that 1,000 is the most likely population size but that there is some probability of many other population sizes, some of which overlap with the range of possible population sizes for 1990. The scientist must decide what estimate for 1990 is sufficiently unlikely to be part of the 1980 distribution to warrant the conclusion that the two are really different.

Scientists are usually cautious about such decisions. If the efficacy of a drug were being tested, it would not contribute to scientific progress if scientists often said a drug did have an effect when it did not. For this reason, the standard risk taken of falsely rejecting the null hypothesis is 1 in 20 (usually phrased $\alpha = 0.05$). In our population example, that means we would reject the null hypothesis for any 1990 population estimate less than the value shown by the dark vertical bar in Figure 8-1a. The area of the 1980 distribution smaller than that value is 5% of the distribution (1/20th).

Now consider our 1990 estimate. The most likely population size is 800. We want to know if the 1990 population is smaller: has our endangered species continued to decline? Presumably, if the population has decreased, a different management action will be taken than if the population had remained stable. Unlike our drug test, there is a good argument that the cost of incorrectly concluding that the population has not declined (Type 2 error) is greater than the cost of incorrectly concluding that there has been a decline (Type 1 error).

The probability of correctly rejecting the null hypothesis is the power of the test. In Figure 8-1a, any 1990 population estimate less than the critical value would lead to rejection of the null hypothesis. Therefore, the area of the 1990 distribution to the left of the vertical bar is the power of the test. Figure 8-1b shows estimates with somewhat lower precision. Power has decreased to a worrisome point: we have only an 11% chance of making a correct decision that the population has declined. If the scientist is willing to take a higher risk of falsely rejecting the null hypothesis, power will be increased. The thin line in Figure 8-1b shows a 1/4 chance ($\alpha = 0.25$) of a Type 1 error, and power is raised from 11% to 50%. If the scientist is willing to make Type 1 errors 50% of the time ($\alpha = 0.5$, reject the null hypothesis for any 1990 estimate less than 1,000), then power is raised to 80%.

Figure 8-2 shows the tradeoff between power and Type 1 errors for estimates of different precision. Population estimates and other data on endangered species are often poor, which means that the failure to reject the null hypothesis (a Type 2 error) often stems from inadequate statistical power rather than any basis in fact. For example, a recent review of the status of 20 cetacean stocks off the coast of California (Barlow et al., 1993) found that only two stocks had population estimates with coefficients of variation of less than 0.3—the median level of precision shown in Figure 8-2—while 11 had population estimates with coefficients of variation between 0.3 and 0. 8 and 7 had population estimates with coefficients of variation even greater than 0.8 (the low level of precision in Figure 8-2). However, those testing hypotheses regarding endangered species often fail to calculate whether they have adequate statistical power (Taylor and Gerodette, 1993).

Box 8-3 Continued

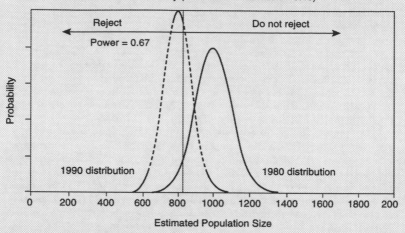

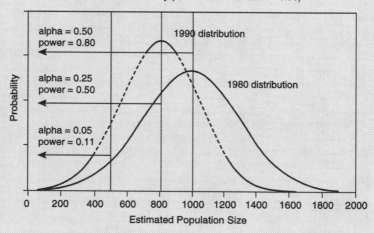

FIGURE 8-1 Power of statistical tests to detect changes in population size. (Top) Comparison of populations with precise size estimates. (Bottom) Comparison of populations with less precise size estimates.

Box continued

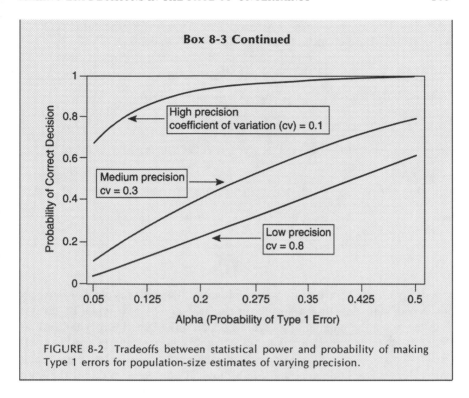

FIGURE 8-2 Tradeoffs between statistical power and probability of making Type 1 errors for population-size estimates of varying precision.

Types of Errors

In the many cases where it is not possible to gather and analyze more information before making a decision, using a formal structure like decision analysis can help managers consider explicitly the ways in which they might be wrong in their predictions (e.g., about whether a particular population is genetically distinct from others or whether loss of a particular habitat area will lead to extinction) and the biological and socioeconomic consequences of being wrong in various ways (such as failing to predict extinction when it will in fact occur, or declaring a population genetically distinct when it is not). Many times the consequences of being wrong are highly asymmetrical; that is, one type of error is much more serious for the species than the converse, and perhaps even irreversible (Box 8-3, Table 8-1). Taylor and Gerrodette (1993) reinforced this point in their discussion of using statistical power analyses to design and evaluate monitoring schemes for endangered species.

Analyzing such situations using the framework of decision analysis helps ensure that all types of errors are considered and that actions chosen are responsive to the likelihood of being wrong and its associated conse-

TABLE 8-1   Consequences of Making Two Types of Statistical Errors
When Evaluating Scientific Data on Endangered Species

Type 1 Error	Type 2 Error
Reject true null hypothesis	Accept false null hypothesis
Claim an effect when none exists	Claim no effect when one exists
Protect species more than necessary	Protect species less than necessary, even lose species
Lose scientific credibility	Lose practical, and scientific credibility
Increase socioeconomic costs more than necessary	Permit activities that should not have been approved

Source: Adapted from Noss (1992).

quences. The ferret and condor capture programs illustrate this point. Managers concerned with delisting a species must be wary of two types of errors: delisting when the population is actually declining (Type I error) and failing to delist when the population is actually stable or increasing (Type II error). (Box 8-3 discusses each type of error in relation to statistical power.) Each type of error has biological and nonbiological consequences, and they are asymmetrical (Table 8-1). The former error has adverse biological consequences for the species—it would be irreversible if the species became extinct—and, perhaps, positive socioeconomic consequences for sectors whose actions are constrained by recovery guidelines. The latter has neutral to positive consequences for the species and possible continued negative socioeconomic consequences. To set acceptable error rates for each type of error, which will in turn determine how many years of monitoring data are required to inform a delisting decision, the magnitude and likelihood of positive and negative consequences using biological and socioeconomic measures must be weighed in a decision analysis framework. These decisions are complicated and consequential enough that unaided intuition cannot always be trusted to do a good job.

   To understand the types of errors that can be made when evaluating scientific data on endangered species and the consequences of these errors, it is helpful to understand the basic logic used when testing hypotheses (Box 8-3). Because statistical tests do not tell us that a hypothesis is true in an absolute sense but merely allow us to control the probability of error in our judgments about whether a hypothesis is true, two types of mistakes are always possible: we might conclude that a hypothesis is false when it is true, or we might conclude that a hypothesis is true when it is false (Box 8-3).

Scientists must try to avoid both types of errors. However, it is impossible to simultaneously minimize the likelihood of making both types of errors. The more we try to avoid making the first type of error, the more likely it is that we will make the second type. Scientists are trained to minimize the probability of making the first error, that is, rejecting a null hypothesis when it is actually true (Box 8-3). This choice is appropriate for advancing scientific knowledge, but it might not be the best for making management decisions. If not examined explicitly, this asymmetric error structure can bias decisions under the act to the detriment of endangered species, especially if they are based on analyses that do not take the asymmetric risk function into account. One situation where this can occur is listing decisions for species where information on population status is limited, a common occurrence. If a statistical analysis is performed, the trigger for listing is rejection of the null hypothesis that the species is not endangered. Typical error rates for such statistical tests of hypotheses keep the likelihood of false rejection low, but at the expense of substantial risk of falsely concluding that a species is not endangered. In the absence of conclusive evidence that a species is in fact endangered, uncertainty about status can result in acceptance of the null hypothesis, whether true or not. This results in an asymmetric risk function for the species (i.e., the probability that the species will not be protected when protection is needed is greater than the probability that the species will be protected unnecessarily), because the null hypothesis is usually that a population has not declined or that a specific action will have no effect (Noss, 1992; Taylor and Gerrodette, 1993). Furthermore, limited data often result in inadequate statistical power. Thus, the null hypothesis of no impact on an endangered species might not be rejected when it should have been (Taylor and Gerrodette 1993). As a result, conservation measures that should be undertaken will not be.

Burden of Proof

In the section above on types of errors that might be made when making decisions under uncertainty, we have shown that the consequences of different types of errors might be asymmetrical, such that it is more important to avoid one type of error than another. We have suggested that ESA decisions should explicitly recognize that the consequences of different types of errors can differ and design decision-making procedures accordingly. An aspect of these decision-making procedures that we would like to emphasize here is the issue of "burden of proof." By this, we mean presumptions about what a party to an ESA decision should have to demonstrate to trigger protective actions. We are concerned that some current procedures may, perhaps inadvertently, bias decision-making in ways that are not intended under the act.

Statistical Errors

In the usual procedures for formulating scientific tests of hypotheses, it is customary to phrase the null hypothesis as the "no effect" case (e.g., a proposed action will not affect the survival of a listed species) and to use confidence levels that limit the probability of falsely rejecting that null hypothesis to a known level (often, 0.05) while permitting much larger probabilities of falsely accepting the null hypothesis. We are concerned that when such statistical procedures are followed in ESA decision-making, they will too often place the burden of proof (for demonstrating a significant effect) on those who want to institute some protective action (usually the FWS or petitioners for listing of a species), without taking into account the practical consequences of falsely concluding that no effect is occurring. This could lead to a systematic bias against species that are candidates for listing or for listed species in need of protective actions.

Cumulative Effects/Thresholds

One situation where this type of problem could arise is when those concerned with species protection suspect that there might be a critical threshold of effect (above which the risk of extinction might increase dramatically) or where cumulative effects might push a species past such a threshold. We have indicated above that our technical ability to predict such thresholds is very limited. If the burden of proof is on those who must show that such a threshold exists (and where it is and just what increase in risk of extinction will occur), there will be few instances in which such a threshold can be demonstrated. As a solution to this problem, we are not advocating that such thresholds simply be assumed unless proven otherwise (which would reverse the burden of proof), but rather that the consequences of each type of error (failing to identify a threshold when one actually exists versus assuming a threshold when one does not exist) be examined to design a decision-making procedure that properly controls the risk of errors, from the point of view of species protection and from the point of view of avoiding unneeded constraints on other interests. In other words, it is advantageous to make the assumptions and their predicted consequences explicit.

Listing Decisions

Another area where we are concerned about asymmetric risk functions for endangered species is in decisions to list them. Lack of information can work against species at risk at the listing stage. Under current conditions, FWS resources for evaluating information on candidate species and for gathering additional information to make a decision are severely limited.

The solution is not simply to reverse the burden of proof and confer protection on all species proposed for listing but to consider explicitly the consequences of both types of error: of listing species that are not really endangered and of failing to list those that really are. Listing decisions are not one of the points in ESA decision making where socioeconomic consequences are to be weighed against species protection, so these need not be part of the equation for determining where the burden of proof should lie for a particular case. However, we acknowledge that with limited time and money for reviewing the eligibility of species for listing, only those species whose situations are known to be the most desperate will receive priority.

Reducing Asymmetry of Risk for Listed Species

In addition to concerns about risks for species at the point of listing, we are also concerned that similar asymmetry of risk functions can occur during decisions regarding protection of already listed species. The usual way of deciding whether there is likely to be a harmful effect is to pose the null hypothesis of no harm and set a low (usually 0.05) rate of error for falsely concluding that there will be harm. This way of framing the question, in combination with limited information on the effects of habitat alteration on listed species, is more likely to deny needed protection than to afford unneeded protection. If the burden of proof were to show that an action would not harm a species rather than to show that it would harm a species, increased protection would result. The importance of shifting the burden of proof this way has been widely recognized, especially in the context of marine conservation issues, and is known as "the precautionary principle" (Cameron and Werksman, 1991; Porter and Brown, 1991; Earll, 1992; Norse, 1993). This principle has already been endorsed in several international legal documents (Porter and Brown, 1991). Recently, the National Marine Fisheries Service (NMFS) explicitly took this kind of asymmetry and the potential for irreversibility into account in deciding to list the anadromous Snake River sockeye salmon (*Oncorhynchus nerka*) as endangered in the face of uncertainty (Waples, in press). The uncertainty concerned whether the anadromous form, which spawns in Redfish Lake, Idaho, was genetically identical to the landlocked form of *O. nerka*, the kokanee, which is common in Redfish Lake. The decision to list[1] was based in part on "the recognition that the consequences of taking the alternate course (i.e., assuming that recent anadromous [fish] in Redfish Lake were derived from

---

[1]*Fed. Reg.* 56:58619, Nov. 20, 1991. Information obtained after the decision indicated that the anadromous form was indeed genetically different from the landlocked form (Waples, in press).

kokanee) and being wrong were irreversible, since the original sockeye salmon gene pool could easily become extinct before the mistake was realized."

Making Tradeoffs Among Competing Objectives

There are always too few resources for the size of the job, and the government has been criticized for allocating funds for species protection without regard for its own stated priorities (Mann and Plummer, 1992). Most people recognize that not everything can be done at once, but these limitations probably would be more acceptable if there were a clearer connection between objectives being pursued and actions taken. Although the language of the ESA suggests that the standards for making decisions about listing, jeopardy, etc., are to be purely scientific, analyses of ESA implementation (e.g., Yaffee, 1982) show clearly that tradeoffs among conflicting objectives must be made in almost every instance. In a few cases, these conflicting objectives and the necessity for balancing them are made explicit in the act and its implementing regulations. For example, in treating critical habitat, the act (Section 4(b)(2)) recognizes that designating critical habitat might have socioeconomic costs and directs those implementing the act to weigh benefits to the listed species against these costs. Similarly, the exemption process specifically directs the exemption committee to ask whether there is an overriding benefit to society from the proposed project that would justify its approval, despite its threat to listed species.

In most cases, however, tradeoffs among competing objectives arise in the course of implementing the act with too few resources, whether financial, human, or natural. For example, there are almost 4,000 candidates for listing (categories 1 and 2), not all of which can be acted on at once. Those responsible for listing decisions must decide which to consider first and which to delay, based on their best judgments about immediacy of threat, distinctiveness of the taxonomic group, etc. Again, structured decision-making techniques can be helpful when deciding on the best use of limited resources. Thibodeau's (1983) application of decision analysis to choose sites for a recovery program is an early application of a structured problem-solving technique to allocate scarce financial resources for conservation. Maguire and Lacy (1990) used decision analysis to help allocate limited zoo space among tiger subspecies in need of captive conservation. To make clear choices among competing objectives and to justify those choices to interested publics, it would be helpful to follow a more explicit framework for evaluating tradeoffs, such as that included in the repertoire of multiattribute decision analysis (Keeney and Raiffa, 1976; Keeney, 1992), which has been applied to several endangered-species problems (Maguire, 1986; Maguire and Serveen, 1992; Ralls and Starfield, 1995).

Among the criteria appropriate for setting priorities among species (or other units of protection), all of which qualify for listing or recovery, are those reflecting how protection of a particular taxon might contribute to maintenance of biological diversity (see Chapter 3). Distinctiveness of various sorts is one measure of this contribution:

Genetic. Does the taxon contain genetic material not represented elsewhere that could provide raw material valuable for adaptation and evolution in future environments?

Phylogenetic. Does the taxon represent a branch of the evolutionary tree of life that has few or no other living members?

Ecological. Does the taxon exhibit an unusual adaptation to its environment, particularly to a rare habitat type (such as vernal pools or hot springs), or does it participate in an unusually close interdependence with other rare (i.e., threatened, endangered, or candidate) species (such as obligate mutualists or parasites), or does it have critical functional roles (i.e., is it a "keystone" species)?

In addition to measures of distinctiveness, other considerations in setting priorities among units of protection include the degree to which conserving that taxon would enhance protection of overall diversity. Higher priority would thus be given to taxa with unusually high levels of genetic diversity, to ecosystems with high levels of endemism, and to taxa whose demise would be likely to precipitate further extinctions of taxa dependent on them. Finally, there are species often referred to as "umbrella species," i.e., they are species whose protection entails the protection of habitats and ecosystems that would confer protection on other (endangered) species. Clearly, if priorities have to be set, an umbrella species should receive a high priority.

FWS (FWS, 1983) and NMFS (NOAA, 1990) have developed hierarchical systems for determining listing, delisting and reclassification, and species-recovery priorities; the NMFS systems are simplified versions of the FWS systems. The FWS system for listing priorities considers the magnitude of the threat to the species, the immediacy of the threat to the species, and the distinctiveness of the species based on its taxonomy, e.g., a monotypic genus is given a higher priority than a species, which in turn is given a higher priority than a subspecies. The system for setting recovery priorities also uses taxonomic level as an indicator of distinctiveness. However, as explained in Chapter 3, taxonomic ranking does not necessarily reflect the same degree of phylogenetic distinctiveness among all groups of organisms.

Resolving Conflicts Among Interest Groups

Parties with a stake in the outcome of ESA decisions include conservationists; developers; other private and public industries; private individuals; academics; local, state, and federal agencies; tribes; and others.   Public participation in ESA decisions is a part of the legislation and implementing regulations in the form of opportunities to petition for listing of species and to offer comments on proposed actions. Negotiated agreements among FWS and interested parties are a common part of ESA implementation through Section 7 provisions for consultation among federal agencies and Section 10 provisions for development of habitat conservation plans by private developers. In Section 7 consultations, FWS is charged with issuing an opinion on whether a proposed action by a federal agency is likely to jeopardize the continued existence of a listed species. In practice, the proposed actions are almost always negotiated informally, with comments by FWS and changes in plans by the developing agency, such that by the time the formal consultation occurs, jeopardy opinions are extremely rare (Houck, 1993).   Although the results might be criticized, the process of negotiating agreement between federal agencies is entirely consistent with directives for constructive cooperation among federal agencies (Cleland, 1991; Wondolleck et al., 1994).

The process of developing habitat conservation plans under Section 10 (Chapter 4; Bean et al., 1991) is also a collaborative, negotiated process between FWS and individuals seeking the incidental take permit. Again, the results of this process often have been criticized.

A combination of methods from decision analysis and dispute resolution can offer disputing parties a way out of the dilemma of how to combine the best scientific analysis with attention to the conflicting values that often are involved in making controversial decisions regarding endangered species. Dispute resolution brings multiple parties together to develop a plan that meets the essential needs of all (Fisher et al., 1991). Decision analysis can help facilitate this process by providing a structure for representing the values and the scientific beliefs that inform each party's positions (Maguire and Boiney, 1994). Decision analysis and dispute resolution direct attention to the objectives and underlying interests that a decision is supposed to reflect. The parts of decision analysis that focus on objectives, priorities, tradeoffs, and criteria can help with this values-structuring part of the analysis (Keeney, 1992). One of the tasks of a collaborative problem-solving process is disentangling disagreements about facts from differences in values. This process is difficult at best, and even harder when there is substantial scientific uncertainty, as is so often the case in ESA decisions. Decision analysis separates each party's view of the facts of the matter from that party's value structure, which is helpful in identifying what each party believes in terms of facts and values. Some of the benefits of decision analysis and dispute resolution can be realized with a qualitative analysis. In other cases, quantitative analysis, such as sensitivity analysis or value of information analysis, can help promote agreement by showing where additional information might help develop a consensus plan and how much additional information might be worth acquiring (Maguire and Boiney, 1994). Such procedures can help ensure that negotiations under Section 7 and Section 10 do a good job of incorporating both science and values.

Implementing Structured Approaches in the Agencies

A call for a more structured approach to ESA decisions using tools such as decision analysis is not necessarily a call for more extensive analysis or research, neither of which could be supported with current resources. Rather, it is a way of making better use of available information to address the problems at hand. Many of the benefits of structured analysis can be realized with a relatively quick and qualitative application of decision and risk concepts (Behn and Vaupel, 1982). Those concepts include explicitly identifying objectives, setting priorities among objectives, and establishing criteria for measuring progress toward those objectives; clearly weighing tradeoffs among conflicting objectives, whether these tradeoffs are forced by the language of the ESA or by limited resources for implementing it; and explicitly considering probabilistic information, making better use of expert opinion, and providing more coherent ways of combining data and opinion to estimate probabilities. All of these can help provide a better connection between the values being pursued under the ESA and the scientific information available to support decision making. In the long run, the decisions will improve, and they will be better justified with reference to both values and science.

It would be possible to provide many of the managers responsible for administering the ESA with the tools to conduct qualitative decision analyses and viability analyses themselves, augmenting the informal methods of analysis they use now. Several short courses and seminars for the Fish and Wildlife Service, U.S. Forest Service, Bureau of Land Management, and state wildlife agencies, all of which have ESA responsibilities, have included decision analysis and population viability analysis as tools for endangered species management.

In some cases, more thorough quantitative analyses will be needed, with consultation from a decision analyst or population biologist. That input could be handled in the same way as subject matter input from species experts when preparing listing documents and recovery plans. Quantitative analysis, such as sensitivity analysis or analysis of the value of information (Maguire, 1986; 1987; Maguire and Servheen, 1992; Ralls and Starfield,

1995), can help build confidence in the robustness of a particular course of action or direct further research to critical parts of a problem.

CONCLUSIONS AND RECOMMENDATIONS

• Major advances in both theory and methods of estimating risk of extinction allow us to base listing and recovery decisions on scientific principles.

- Many previous ESA decisions did not meet the guidelines suggested by current scientific thinking, listing species as endangered only after populations had dropped to the point where the risk of extinction was very high and proposing recovery goals that left the species still at high risk of extinction.

- Where natural history and demographic data are available, analytical and simulation models can be used to provide quantitative estimates of risks of extinction.

- General results from these extinction models have been used to develop rating systems based on objective criteria (such as population size, number of populations, and other demographic and environmental characteristics) to categorize species according to relative risk of extinction. Rating systems for use in situations where detailed data are not available should be developed and tested with simulation and observational methods.

- Because current extinction models do not consider the interactions of all factors promoting extinction, estimates of extinction risk might underestimate the true risk.

• Setting levels of risk to trigger listing and recovery decisions entails scientific and public policy considerations.

- We can find no scientific basis for setting different levels of risk for different taxonomic groups, such as plants and animals, or for public versus private actions that might affect listed species. However, it is critical to understand that achieving the same *biological* risks for listed species might well entail different management policies on public and private lands, because public and private entities behave differently from each other. No implementation of the Endangered Species Act can be fully successful without recognition of these differences.

- To the degree that they can be quantified, the levels of risk associated with endangered status should be higher than those for threatened status. Once a species no longer qualifies as threatened, it should be considered recovered and delisted.

- Levels of risk to trigger ESA decisions must be framed as a probability of extinction during a specified period (e.g., $x\%$ probability of extinction over the next y years). Although some crises might call for short time horizons (on the order of tens of years), ordinarily it will be necessary to view extinction over longer periods (on the order of hundreds of years), so that short-term solutions do not create long-term problems.

- The selection of particular degrees of risk associated with particular periods to trigger ESA decisions reflects scientific knowledge and societal values.

- When implementing the ESA with limited resources, it will probably be necessary to allocate effort among species, all of which qualify for protection according to the risk level that has been adopted. Scientific considerations, such as whether a species or its habitat possesses unusually distinctive attributes or whether protection of a taxon would confer protection on other candidate taxa or their habitats, should help set priorities for action.

• There will always be uncertainty in the estimates of risk used to trigger decisions under the ESA, requiring policies and processes for making decisions with incomplete and uncertain data. Making decisions under uncertainty poses the possibility of errors of various types, such as delisting when a species has not actually recovered or listing when a species is not really endangered or threatened. For a variety of statistical reasons, including those pertaining to availability of data, protection would be more likely if the burden of proof were to show that a proposed action would not harm a listed species rather than to show that it would.

• Because ESA decisions are often difficult and controversial, the procedures used to make them should be explicit and well documented. Structured methods can improve the substance of these decisions and the justification for them. Structured methods can be particularly appropriate to ESA decisions when

- scientific risk assessments and societal values must be integrated;
- tradeoffs among conflicting objectives must be made or negotiations among disputing parties must be conducted;
- the costs and benefits of delaying decisions while gathering additional information to reduce uncertainty must be weighed; and
- empirical data are lacking and information derived from expert opinion should be used.

REFERENCES

Barlow, J., J. Sisson, and S.B. Reilly. 1993. Status of California Cetacean Stocks: A Summary of the Workshop held March 31 to April 12. Administrative Report LJ-93-20. National Marine Fisheries Service, Southwest Fisheries Science Center, La Jolla, Calif.

Bart, J., and D. Robson. In Press. Design of a monitoring program for northern spotted owls. In Proceedings of the Symposium on Monitoring Bird Population Trends by Point Counts, C.J. Ralph, J.R. Sauer, and S. Droege, technical coordinators. USDA Gen. Tech. Rep. PSW-000. U.S. Department of Agriculture Forest Service, Pacific Southwest Research Station, Albany, Calif.

Bean, M.J., S.G. Fitzgerald, and M.A. O'Connell. 1991. Reconciling Conflicts Under the Endangered Species Act: The Habitat Conservation Planning Experience. World Wildlife Fund, Washington, D.C.

Beanlands, G.E., W.J. Erckmann, G.H. Orians, J. O'Riordan, D. Policansky, M.H. Sadar, and B. Sadler. 1986. Cumulative Environmental Effects: A Binational Perspective. Canadian Environmental Assessment Council, Ottawa, Ontario, and the National Research Council, Washington, D.C.

Behn, R.D. and J.W. Vaupel. 1982. Quick Analysis for Busy Decision Makers. New York: Basic Books.

Brownell, R.L., Jr., K. Ralls, and W.F. Perrin. 1989. The plight of the "forgotten" whales. Oceanus 32(1):5-11.

Cameron, J., and J.D. Werksman. 1991. The Precautionary Principle: A Policy for Action in the Face of Uncertainty. Centre for International Environmental Law, London.

Cleland, J.C. 1991. Application of Alternative Dispute Resolution to Endangered Species Act Interagency Consultations. Master of Environmental Management Project, School of Forestry and Environmental Studies, Duke University, Durham, N.C.

Clemen, R.T. 1991. Making Hard Decisions. An Introduction to Decision Analysis. Boston, Mass.: PWS-Kent.

Donovan, G.P., ed. 1989. The Comprehensive Assessement of Whale Stocks: The Early Years. International Whaling Commission, Cambridge, U.K.

Earll, R.C. 1992. Common sense and the precautionary Principle—An environmentalist's perspective. Mar. Pollut. Bull. 24:182-186.

Fisher, R., W. Ury, and B. Patton. 1991. Getting to Yes. 2nd. ed. New York: Penguin Books.

FWS (U.S. Fish and Wildlife Service). 1983. Endangered and threatened species listing and recovery priority guidelines. Fed. Reg. 48(184):43098-43105.

Gnam, R. 1993. Comments invited on species' risk. BioScience 43:430.

Goodman, D. 1993. Scientific standards for endangered species management. Appendix 1 in Research on Methods of Biodiversity Management. Annual Report No. 1. Cooperative Agreement No. CR-8200-8601. U.S. Environmental Protection Agency, Office of Research and Development, Washington, D.C.

Hogarth, R. 1980. Judgment and Choice: The Psychology of Decisions. Chichester, U.K.: John Wiley & Sons.

Houck, O.A. 1993. The Endangered Species Act and its implementation by the U.S. Departments of Interior and Commerce. Univ. Colo. Law Rev. 64(2):277-370.

Kahneman, D., P. Slovic, and A. Tversky. 1982. Judgement Under Uncertainty: Heuristics and Biases. Cambridge, U.K.: Cambridge University Press.

Keeney, R.L. 1992. Value-Focused Thinking. Cambridge, Mass.: Harvard University Press.

Keeney, R.L., and H. Raiffa. 1976. Decisions with Multiple Objectives: Preferences and Value Tradeoffs. New York: John Wiley & Sons.

Lande, R. 1988. Demographic models of the northern spotted owl (*Strix occidentalis caurina*). Oecologia 75:601-607.

Mace, G., N. Collar, J. Cooke, K. Gaston, J. Ginsberg, N. Leader Williams, M. Maunder, and E.J. Milner-Gulland. 1992. The development of new criteria for listing species on the IUCN Red List. Species 19:16-22.

Mace, G.M., and R. Lande. 1991. Assessing extinction threats: Toward a reevaluation of IUCN threatened species categories. Conserv. Biol. 5:148-157.

Maguire, L.A. 1986. Using decision analysis to manage endangered species populations. J. Environ. Manage. 22:345-360.

Maguire, L.A. 1987. Decision analysis: A tool for tiger conservation and management. Pp. 75-486 in Tigers of the World, R.L. Tilson and U.S. Seal, eds. Park Ridge, N.J.: Noyes.

Maguire, L.A. 1989. Managing black-footed ferret populations under uncertainty: An analysis of capture and release decisions. Pp. 268-292 in Conservation Biology and the Black-footed Ferret. U.S. Seal, M. Bogan, T. Thorne, and S.F. Anderson, eds. New Haven, Conn.: Yale University Press.

Maguire, L.A. and L.G. Boiney. 1994. Resolving environmental disputes: A framework incorporating decision analysis and dispute resolution techniques. J. Environ. Manage. 42:31-48.

Maguire, L.A. and R.C. Lacy. 1990. Allocating scarce resources for conservation of endangered species: Partitioning zoo space for tigers. Conserv. Biol. 4:157-166.

Maguire, L.A., and C. Servheen. 1992. Integrating biological and sociological concerns in endangered species management: Augmentation of grizzly bear populations. Conserv. Biol. 6:426-434.

Mann, C.C. and M.L. Plummer. 1992. The butterfly problem. Atlantic Mon. 269(1):47-70.

Menges, E.S. 1990. Population viability analysis for an endangered plant. Conserv. Biol. 4:52-62.

NOAA (U.S. National Oceanic and Atmospheric Administration). 1990. Endangered and threatened species; listing and recovery priority guidelines. Fed. Reg. 55(116):24296-24298.

Norse, E.A., ed. 1993. Global Marine Biological Diversity: A Strategy for Building Conservation into Decision Making. Washington, D.C.: Island Press. 383 pp.

Noss, R.F. 1992. Biodiversity: Many scales and many concerns. Pp. 17-22 in Proceedings of the Symposium on Biodiversity of Northwestern California, H.F. Kerner, ed., Oct. 28-30, Santa Rosa, Calif.

NRC (National Research Council). 1992a. The Scientific Bases for the Preservation of the Hawaiian Crow. Washington, D.C.: National Academy Press.

NRC (National Research Council). 1992b. Decline of the Sea Turtles: Causes and Prevention. Washington, D.C.: National Academy Press.

Porter, G., and J.W. Brown. 1991. Global Environmental Politics. Boulder, Colo.: Westview.

Raiffa, H. 1968. Decision Analysis: Introductory Lectures on Choices under Uncertainty. Reading, Mass.: Addison-Wesley.

Ralls, K., and J.D. Ballou. 1992. Managing genetic diversity in captive breeding and reintroduction programs. Trans. North Am. Wildl. Nat. Resour. Conf. 57:263-282.

Ralls, K., and A.M. Starfield. 1995. Two decision analysis methods for conservation problems. Conserv. Biol. 9:175-181.

Reilly, S. 1992. Population biology and status of eastern Pacific gray whales: Recent developments. Pp. 1062-1074 in Wildlife 2001: Populations, D.R. McCullough and R.H. Barrett, eds. London: Elsevier Applied Science.

Saaty, T.L. 1990. How to make a decision: The analytic hierarchy process. Eur. J. Oper. Res. 48:9-26.

Shaffer, M.L. 1981. Minimum viable population sizes for species conservation. BioScience 31:131-134.

Taylor, B.L. and T. Gerrodette. 1993. The uses of statistical power in conservation biology: The vaquita and northern spotted owl. Conserv. Biol. 7:489-500.

Tear, T.H., J.M. Scott, P.H. Hayward, and B. Griffith. 1993. Status and prospects for success of the Endangered Species Act: A look at recovery plans. Science 262:976-977.

Thibodeau, F.R. 1983. Endangered species: Deciding which species to save. Environ. Manage. 7:101-107.

von Winterfeldt, D., and W. Edwards. 1986. Decision Analysis and Behavioral Research. Cambridge, U.K.: Cambridge University Press

Waples, R.S. In press. Evolutionarily Significant Units and the Conservation of Biological Diversity Under the Endangered Species Act. In Evolution and the Aquatic Ecosystem, J.L. Nielsen and D.A. Powers, eds. American Fisheries Society, Bethesda, Md.

Wilcove, D.S., M. McMillan, and K.C. Winston. 1993. What exactly is an endangered species? An analysis of the endangered species list, 1985-1991. Conserv. Biol. 7:87-93.

Wondolleck, J.M. , S.L. Yaffee, and J.E. Crowfoot. 1994. Applying the principles of alternative dispute resolution to endangered species conservation. In Improving Endangered Species Conservation: Reviewing the Experience and Learning the Lessons, T. Clark and A. Clarke, eds. Covelo, Calif.: Island Press.

Yaffee, S.L. 1982. Prohibitive Policy: Implementing the Federal Endangered Species Act. Cambridge, Mass.: MIT Press.

9

Areas of Scientific Uncertainty

The main purpose of the Endangered Species Act is to provide protection for species with an uncertain future, and uncertainty permeates all decisions made under the act. This chapter focuses on the major areas of *scientific* uncertainty that exist with respect to applications of the ESA. The emphasis is on uncertainties that could be resolved with further research, as opposed to intrinsic uncertainties in species survival. Even in the best of possible worlds, with perfect data and valid estimation and evaluative procedures, there is always a probabilistic element to any assessment of risk. Nonetheless, the committee concludes that none of the scientific uncertainties discussed below is great enough to make the ESA unworkable.

ECOSYSTEM-BASED PROTECTION

A stated purpose of the ESA is "to provide a means whereby the ecosystems upon which endangered species and threatened species depend may be conserved . . . ." The means to this end is the listing of individual species. The major threat to most species is loss of habitat, and therefore ecosystem protection is of paramount importance to the overall preservation of species. Because the ESA requires that critical habitat be designated at the time of listing, listing a species has the potential to protect ecosystems and their unlisted components as well. However, this approach can be effective only if habitat protection is pursued rigorously.

Less clear is whether listing species, as opposed to a broader based policy of listing ecosystems, is the best means of achieving this goal. Protecting ecosystems is probably the only way to ensure the long-term survival of large numbers of species, but the best way to achieve such protection is uncertain.

Ecosystem Management

Species are relatively easy to identify. Ecosystems are difficult to define and certainly more difficult to manage (e.g., see Franklin, 1993; Irwin and Wigley, 1993; Naiman et al., 1993; Wilcove, 1993). For example, a lake ecosystem can be defined by the boundaries of its shoreline or by its shoreline and the terrestrial watershed on which it is critically dependent. Ecosystem protection is a fairly new concept, and policy for implementing it is untested. Nonetheless, it appears to the committee that enough is known to be helpful. Indeed, several federal agencies have expressed their desire to adopt ecosystem-management approaches, and some have developed task-forces to develop those approaches.

Definitions of ecosystem management tend to fall into two major categories. The first—some concept of management to achieve various ecosystem goals—is the more difficult to implement. The second category is the idea of keeping other ecosystem components and processes in mind when managing a particular part of an ecosystem. This would mean, for example, that one would keep in mind the needs of marine mammals and birds when harvesting fish; one would keep in mind aquatic ecosystems when managing adjacent uplands (whether for forestry, agriculture, grazing, recreation, development, or any other goal); and one would keep various ecosystem processes and components in mind when managing for protection of endangered species. The second category is already being developed or practiced by many people in federal and state agencies (e.g., LaRoe, 1993; Quigley and McDonald, 1993), and it has the potential to help protect endangered species, to help protect the ecosystems they depend on, and to help reduce social and economic disruption and conflict. Therefore, despite the need for more knowledge, experience, management tools and, in some cases, social acceptance, ecosystem management offers promise.

INADEQUATE KNOWLEDGE OF SPECIES AND THEIR ROLES IN ECOSYSTEMS

The Endangered Species Act has been applied almost exclusively to vertebrates, invertebrates, and vascular plants. For small or inconspicuous organisms, a large fraction of the biota probably has not been classified. Furthermore, new species even of conspicuous taxa are still being discovered (Wilson, 1988). Obviously, organisms that have not been identified cannot be evaluated and protected if warranted.

A fundamental characteristic of an evolutionary unit (see Chapter 3) is that an EU is distinct from other units. Whether a population segment in the wild is distinct or part of a larger genetic entity is often unclear because historical and current levels of gene flow are unknown. Furthermore, evolutionary change is dynamic, and tests for distinctiveness are most difficult to apply when populations are diverging into independent populations.

On scales of tens to thousands of years, most species expand and contract in number and geographic distribution in response to environmental change, and they evolve. But we do not know how many species can be lost before an ecosystem itself collapses. The roles of most species in most ecosystems remain unknown for described and undescribed species. It is known, however, that complex ecosystems can exhibit sudden changes in state once a threshold level of stress has been exceeded (Begon et al., 1986). Thus, we dare not lose sight of the fact that species currently kept rare by natural or human-induced factors play or could play central roles in the biosphere in the future.

ESTIMATION OF THE RISK OF EXTINCTION

Current Limitations of Existing Theory

Nearly all of what is now a substantial body of theory for predicting the risk of extinction has been developed since the ESA was initiated in 1973. The major accomplishment of the theory to date is the identification of ways in which expected times to extinction scale with population size when a single factor is the dominant source of risk (e.g., demographic versus environmental stochasticity versus episodic catastrophes). Even this level of work has been very difficult, and numerous assumptions have been made to obtain reasonably simple analytical solutions. For example, almost all existing analytical models ignore the age, spatial, and genetic structures that are inherent in most natural populations.

Although they have heuristic value, unifactorial models of extinction have limited utility in the real world of risk assessment for the simple reason that small populations always are confronted simultaneously with threats from demographic, environmental, and genetic stochasticity. Factors that could reduce population size can be highly synergistic, and each one might spawn further stochasticity in the other. Such interactions can lead to greatly elevated risks of extinction. Thus, unifactorial models might provide us only with *the lower limits* of the risk of extinction. Of course, no model can ever be expected to produce perfect estimates of risk. However, from the standpoint of species protection, upwardly, rather than downwardly, biased estimates of the risk of extinction are preferred, so that errors in risk assessment would tend to be on the side of species.

Intrinsic Limits of Extinction Models

Biological models that jointly incorporate demographic and environmental (spatial and temporal) variation, age structure, and genetics can be analyzed by computer simulation. However, predictions emanating from these models will always be subject to uncertainty. Most notably, some aspects of the structure of the model (e.g., mode of density dependence, temporal and spatial patterns of environmental variation, and frequency and magnitude of catastrophes) will almost always be in doubt. Even for rare cases in which the essential information is available, its relevance to predictive models can be limited. For example, fundamental features of population structure and dynamics of a species in jeopardy because of environmental change might be altered in unanticipated ways. To a certain extent, those types of uncertainty can be dealt with by using a model structure and conservative enough parameter estimates that the predicted risk of extinction will most likely be an overestimate. In addition, evaluation of the sensitivity of a model's predictions to variation in its parameters can be used to identify the features of a population for which accurate estimates are most critical to the decision process. These sensitivity analyses should be conducted routinely, and the results should be used to direct future research.

LACK OF BASIC INFORMATION

Whether explicitly or implicitly, all decisions concerning rare, threatened, or endangered species are based on assessments that have at least some quantitative basis, even if that basis is not explicit. Yet, critical data to make informed decisions on proposals for listing, to designate critical habitat, and to develop recovery and management plans are usually lacking. Our biological understanding of many rare, threatened, or endangered species does not extend far beyond a taxonomic description and a coarse geographic distribution. That lack of data should not be the basis for failure to list a species if other information is available to indicate that listing is otherwise warranted. The act calls for the use of the best scientific data available in the decision-making process. It does not, and should not, require that all desirable data be available at the time of listing.

Dynamics of Natural Populations

Recovery plans often set goals based on target population sizes. Equally important is the need to stabilize the mean population density. One of the largest gaps in our knowledge of the population biology of most species concerns the natural temporal and spatial variation that exists in key demographic factors. That information is critical to evaluating the risk of extinction, regardless of the mean population size.

Systematics

Protection of ecosystems is becoming recognized as an attractive option for conserving biological diversity. However, ecosystems are composed of species and populations, and those components of ecosystem structure must be understood. Yet the vast majority of species in the United States are unknown and unnamed. Even for many of the named species, virtually nothing is known of their geographic ranges, population structure, demography, ecology, or practically any aspect of their biology. We have only the roughest of estimates of how many species of organisms reside in this country. A recent NRC report on the National Biological Survey (now the National Biological Service) (NRC, 1993a) recommended a commitment to a detailed study of a significant portion of our biota. Although some of the more visible vertebrate groups, such as birds and mammals, are well known, many plant groups and most invertebrates, except for some commercially important ones, remain virtually unstudied. Some large ecologically important groups have few or no systematists studying them. Any realistic attempt to provide even a basic inventory of our biota will need significant new resources for training and supporting systematic biologists. Wise understanding, management, and conservation of our biota need a much better picture of what organisms inhabit our country.

Do Minimum Viable Population Sizes Exist?

A popular heuristic concept in conservation biology is that of a minimum viable population size (MVP), i.e., a threshold population size below which rapid extinction is virtually guaranteed. Should MVPs exist in reality, numerical guides to them would be useful as listing criteria. At this point, there is little compelling evidence that general guidelines can be made in this regard. Certainly, small populations are more vulnerable to extinction than large ones, but it remains to be seen whether there is some critical population size below which the vulnerability to extinction increases suddenly. It is perhaps more useful to estimate extinction probabilities as a function of time for different population sizes than to identify some specific MVP, as discussed in Chapter 7.

THE PROTECTION OF GENETIC DIVERSITY

In previous chapters, we described the importance for the survival of species of maintaining genetic diversity for adaptive characters within and between populations. All species are now, and perhaps always have been, confronted with a globally changing environment. Many rare species have the additional burden of being confined to habitats that are changing rapidly in response to local human activity. Although all species have evolved behavioral and physiological mechanisms for coping with environmental change, the range of environments within which such homeostatic mechanisms are operative is normally confined to the conditions experienced over long periods. For species encountering entirely new environmental conditions, evolutionary flexibility is essential for long-term survival. Thus, the preservation of diversity at the species level is intrinsically dependent on the maintenance of genetic diversity within species. The difficulty lies in the identification and quantification of this genetic diversity.

Uncertainty Regarding Future Adaptive Challenges to Species

If information on all quantitative-trait variation could be obtained for an endangered species, it would still be difficult to identify which characters should be evaluated, because we would be unsure of the selective challenges that would confront species in the future or the characters that will contribute to adaptive change. With this uncertainty, the best strategy for the maintenance of genetic diversity within species is the implementation of protection programs that are likely to maximize genetic variation for all characters. Programs designed to maximize effective population size will naturally maximize the expected amount of genetic variation as well. Because individual genomes are mutable, populations that are devoid of useful quantitative-genetic variation can replenish that variation over tens of generations and should not be ruled out as viable evolutionary lineages.

FEASIBLE MANAGEMENT STRATEGIES

Perhaps the paramount challenge to future managers of endangered species concerns the degree to which management and recovery plans can be developed within a framework that incorporates a range of continuing human activities. Numerous issues remain unresolved, such as the design of reserves, reconstruction of habitat, the usefulness of captive breeding and supplementation programs, and the effects of environmental change.

The Spatial Structure of Reserves

A major challenge for conservation biology is the need to develop methods for ascertaining optimal strategies for moving species toward recovery goals when resources and critical habitat are in limited supply, as they always are. The spatial arrangement of habitats can have substantial effects on the persistence of metapopulations, but our understanding of even the most basic issues is still undeveloped (NRC, 1993b). Analogous to the concept of minimum viable population size, there might be a threshold number of subpopulations or a threshold degree of isolation beyond which a metapopulation becomes highly vulnerable to extinction, although this would certainly be expected to vary from species to species, depending on their biological features.

Corridors and Edge Effects

In principle, corridors between local demes can allow metapopulations of the demes to serve as buffers from extinction in a stochastically varying environment. However, because of their large edge effects, corridors often contain inhospitable habitat through which migration is risky. Consequently, corridors can be sinks as well as sources of individuals in a metapopulation context. Attempts to evaluate whether management of a species should involve a few very large reserves versus many smaller ones will be shortsighted if they do not take into account the demographic consequences of corridors (NRC, 1993b).

Fragmentation of habitat, in general, is a particularly serious area of uncertainty. Because ecosystem structure develops over several hundreds to thousands of years, several human generations could pass before the full consequences of habitat fragmentation and the resulting edge effects were revealed.

Reconstruction of Habitat

In habitat conservation plans involving mitigation, proposals by developers to reconstitute ecosystems at alternative sites are becoming increasingly common. Careful management of species whose biology is well understood can lead to their protection in altered environments. However, development of complex communities for listed species must be approached cautiously, because we often are dealing in theory rather than proven ability. Reconstituted ecosystems can have very different internal and external interactions than their predecessors (NRC, 1992). As a consequence, maintenance of such ecosystems might require long-term and, perhaps at times, intensive management. Ecosystems, like species, evolve over time and space as their component species wax and wane. Artificial manipulation (management) might therefore be necessary if we are to focus on a particular listed species or species set as a target for management.

Consequences of Captive Breeding and Supplementation

From a genetic perspective, a fundamental issue for which we have almost no empirical information is the degree to which semi-isolated populations develop genomic incompatibilities, which upon crossing, would be exhibited as reduced fitness of the offspring. This issue is becoming increasingly important as recovery plans incorporate captive breeding, supplementation, and sometimes, hybridization procedures into management policies.

Global Environmental Change

In applications of the ESA, the major focus on species protection has been on local issues, such as dam and road building, logging and mining, grazing, and housing development. However, evidence suggests that human activity is causing global changes in temperature and the chemical composition of the atmosphere (Abrahamson, 1989; Kareiva et al., 1993). Even before humans had the capacity to cause environmental changes at larger than local scales, regional and global environments were changing; indeed they have changed as long as life has been on earth (see Chapter 2). Those types of changes, particularly when combined with habitat fragmentation, could pose major threats to rare and sensitive species. Policies for managing biodiversity will be short-sighted if they are developed in a setting that does not consider the implications of global environmental change.

VALUING RARITY

Many uncertainties in economics (defined broadly as the science of human choice and valuation) relate to the Endangered Species Act. One of the largest of these concerns the valuation of rarity.

Valuation is an enormously controversial topic. Some cognitive psychologists argue that strong environmental values are not represented in monetary form in people's mental models (e.g., Gregory et al., 1993). Tversky et al. (1988) argued that the way people rank and order items depends on the measure used. Applied to the case of endangered species, it means that people might put a higher dollar value on one species than another but reverse the ordering if asked to decide which species should be preserved to make the greatest contribution to genetic diversity. Many issues are contentious between the fields of economics and ecology. Indeed, some economists and philosophers have doubted the ability of economics to solve this question because of the diversity of attributes being evaluated, the lack of information available, and especially the difficulty of including moral and long-term considerations in the valuation (e.g., Norton, 1988; Norgaard, 1988). Others have pointed out the usefulness of having some sort of balanced and complete economic analysis, even if it includes only short-term considerations (e.g., Randall, 1988). This discussion is limited to the economic perspective.

Value does not inhere in objects. The attribution of values to objects by individuals is motivated by cultural and religious underpinnings, but our tastes and preferences are also influenced by more transitory forces of television, advertising, and the print media. Therefore, explaining taste is presently very much an exploratory enterprise.

Rare things often are valued because ownership is a conspicuous way of displaying superiority. So some are willing to pay a great deal to possess a private good that few others can afford to have. Unfortunately, no research has established the relationship between rarity and dollar value. But rarity and great value are not the monopoly of private goods. Many feel great exaltation in front of Michelangelo's Pietà; looking at one of Christo's wraps; or observing natural wonders, such as the great migrations of zebras, wildebeests, and other animals between Kenya and Tanzania, or tens of thousands of migratory birds taking flight from a lake. The economic analog of these ideas is a willingness to pay if necessary to enjoy these experiences rather than go without one or more of them. Great economic value can arise from great quantity and is not limited to things quantitatively scarce, as long as qualitative attributes are acknowledged.

Goods and services have high economic value when they are economically scarce, i.e., when the demand for them is large relative to supply. A key element in explaining whether consumers place a high or low value on something is the availability of substitutes. The destruction of something we like enormously is not so bad if we can easily find a substitute. Few substitutes is a necessary but not sufficient condition for high value. (For example, rare fatal diseases are not particularly valuable.) Not all rare things are valuable.

Endangered species are, by definition, rare (or nearly so), but quantitative rareness is not a sufficient attribute to conclude that any and all endangered species have great economic value. Wilson (1988) and others effectively have heightened public awareness of the accelerated pace of species extinction—in recent years, three per hour in the rain forests alone, according to Wilson's latest estimate (Wilson, 1992). Yet we are complacent with that knowledge and with knowledge of threats to tropical rain forests and other hotspot ecosystems around the world. In the final analysis, allocated funds reveal how valuable the citizenry thinks endangered species are and how much it is willing to give up other things to have greater preservation activities. Congress annually appropriates funds for the Office of Endangered Species in the United States that are not adequate to list more than a small fraction of the candidate species or to pay for more than a fraction of the possible recovery plans for all endangered species. U.S. voters and their representatives in state and national legislatures have yet to demonstrate enthusiasm in support (or willingness to make sacrifices for) of species preservation despite the belief of many scholars and researchers that the pace of extinction is too rapid.

Preservation efforts might be facilitated if estimates of economic value of rare and endangered species were available. Estimates of value are elusive because of the nature of the benefits (Brown, 1990); with few exceptions, credible estimates do not exist.

Many suggest that a species is worth preserving if it yields products of commercial worth. It is easy to find specific plants with great commercial value, such as the rosy periwinkle (*Catharanthus roseus*) which has been used in a cure for acute lymphocytic leukemia and Hodgkin's disease. But generalization of an estimate of economic value to all species is more problematical, particularly since the chance of finding a product of economic value is so small, perhaps on the order of 1 in 10,000 (Aylward et al., 1993). And preservation is costly and requires tradeoffs. It is not surprising that companies are reluctant to make privileged information about specific costs and revenues public, so data do not exist to estimate the commercial economic value of genetic resources in general and any species in particular. Preserving for commercial value is not a good strategy, unless endangered species can be ranked according to the chance of successful discovery of commercial products and expected value if successful.

Even if the expected economic return could be accurately predicted, estimating the commercial value of species correctly will result in a systematic underestimate of species' value to society. If the species are common property, owned by none, then others can use them directly or indirectly for competitive commercial gain, so that value of any template or product developed from a species is devalued by the first discoverer who knows that a successful rival cannot be far behind. Patents are an imperfect protection. Removing common property status is tantamount to privatization, in which case the private "owners," apart from exceptional cases, act as monopolists and exploit their monopoly power to the detriment of social welfare.

Many species are valuable because they provide either food or recreation directly to the consumer. However, this direct consumptive value is an excluded source of value for rare species. The benefits we derive from species as goods—either commercial or consumptive—are relatively easy to value compared with another type of value that derives from the services that species perform within the ecosystems that contain them. Such services include the maintenance of fertile soil and water, control over the composition of the atmosphere, and regulation of the climate and the hydrological cycle (including flood control), and pest control. These major benefits to the human economy and to human well-being are called *ecosystem services* by ecologists, but that phrase masks the important roles that individual species or groups of species play in providing those services. For example, some species of microorganisms (denitrifying bacteria) convert nitrate in soil into a gas, nitrous oxide, that plays an important role in regulating the concentration of atmospheric ozone. Difficulty in valuing the roles of individual species in providing these services arise from uncertainty over the actual economic value of fertile soil, clean water, and other ecosystem-derived benefits. Indeed, Norton (1988) equated that value to "the summed value of all the GNPs of all countries from now until the end of the world."

Another part of a species' value is called *non-use* value. We can derive value from species by knowing they exist today—for example, the value of viewing and photographing them. An illustrative study of these values for elk, bighorn sheep, and grizzly bear was reported by Schulze et al. (1981). Many studies document that a substantial fraction of species values arises if we can be assured that they will be around in the future for subsequent generations to enjoy.

The literature on the estimation of the non-use value of species is very modest. All studies estimating non-use value use a contingent valuation method, discussed below. The value of preserving the whooping crane population at the Arkansas National Wildlife Refuge in Texas for viewers and non-viewers has been estimated by Stoll and Johnson (1984) and Bowker and Stoll (1988). Hageman (1985) valued blue whales, bottlenose dolphins, California sea otters and northern elephant seals. Brown and Henry (1989) estimated the value of preserving elephants in Kenya. Boyle and Bishop (1987) estimated the value of preserving the striped shiner, a Wisconsin endangered species. Boyle and Bishop (1986) also have estimated the existence value of eagles by focusing, in part, on whether respondents view eagles or not. Brown et al. (1994) estimated the value of the northern spotted owl, as have Hagen et al. (1991).

The valuation of non-use by economists is new and controversial. By its very nature, such valuation is not founded on behavioral observations, which is the source of controversy. Non-use values cannot be observed from organized markets. The research method is contingent valuation (Cummings et al., 1986; Mitchell and Carson, 1988). It involves the design of a survey that elicits dollar values that, for example, represent a respondent's willingness to pay for the preservation of one or more rare species. Critics argue that the values estimated from contingent value surveys are hypothetical and lack credibility. Advocates rebut that socioeconomic factors considered to be determinants of value have the right sign (i.e., they are positive when expected to be and negative when expected to be) and are statistically significant in well-designed studies and that more than 1,000 contingent valuation studies have been done in more than 40 countries.

Controversy over the method of contingent valuation was sparked by the *Exxon Valdez* oil spill, because the regulations call for such studies. The National Oceanic and Atmospheric Administration, a federal trustee for natural resources injured by oil spills, created a panel composed of Nobel Prize winners in economics and other experts. The panel approved of the contingent valuation method, providing certain criteria were met (NOAA Panel on Contingent Evaluation, 1993).

The case on economic grounds for preserving endangered species depends crucially on the magnitude of non-use values for species. Although we may believe that any or all endangered species are too valuable to sacrifice, there is an inadequate scientific basis to demonstrate whether citizens are willing to make the sacrifices necessary to save all endangered species in this country now and in the foreseeable future. In addition, economic analyses are less effective when assessing long-term values than short-term ones. The expected short-term use value in monetary terms of preserving many of the tens of millions of extant species is likely to be small relative to the short-term real costs of saving them (Brown, 1990; Gregory et al., 1993), especially if externalities such as ecosystem goods and services are not factored into the analysis. In part because of uncertainties in biological knowledge, the long-term costs and benefits of protecting endangered species and their ecosystems is poorly known.

In our world of limited resources, the harsh fact is that we must give to get. In the absence of scientific facts, belief, not science, defends the view that endangered species are more economically valuable to citizens of the United States than the value of resources it will take to save them. However, many policy decisions concerning public goods are made without compelling economic arguments. It has also been argued that economic and ecological values are consistent with each other and that this consistency should be recognized by policy makers (e.g., Ashworth, 1995), so inasmuch as preserving species is related to preserving ecosystem functioning, preserving species should lead to an enhancement of both ecological and economic values.

REFERENCES

Abrahamson, D.E. 1989. The Challenge of Global Warming. Washington, D.C.: Island Press.
Ashworth, W. 1995. The Economy of Nature: Rethinking the Connections Between Ecology and Economics. New York: Houghton-Mifflin.
Aylward, B.A., J. Echerria, L. Fendt, and E. Barbier. 1993. The Economic Value of Species

Information and Its Role in Biodiversity Conservation: Case Studies of Costa Rica's National Biodiversity Institute and Pharmaceutical Prospective. London Environmental Economics Centre, London, U.K.

Begon, M., J.L. Harper, and C.R. Townsend. 1986. Ecology: Individuals, Populations, and Communities. Sunderland, Mass.: Sinauer Associates.

Bowker, J.M., and J.R. Stoll. 1988. Use of dichotomous choice nonmarket methods to value the whooping crane resource. Am. J. Agr. Econ. 70 (2):372-381.

Boyle, K.J., and R.C. Bishop. 1986. The economic valuation of endangered species of wildlife. Trans. North Am. Wildl. Nat. Resour. Conf. 51. p. 153-161.

Boyle, K.H., and R.C. Bishop. 1987. Toward total valuation of Great Lakes fishery resources. Wat. Resour. Res. 5:943-950.

Brown, G.M., Jr. 1990. Valuation of genetic resources. Pp. 203-229 in The Preservation and Valuation of Biological Resources, G.H. Orians, G.M. Brown, Jr., W.E. Kunin, and J.E. Swierbinski, eds. Seattle, Wash.: University of Washington Press.

Brown, G.M., Jr., and W. Henry. 1989. The Economic Value of Elephants. International Institute for Environment and Development. LEEC Paper 89-12. London Environmental Economics Centre, London, U.K.

Brown, G.M., Jr., D. Layton, and J. Lazo. 1994. Valuing Habitat and Endangered Species. Institute for Economic Research, University of Washington, Seattle, Wash.

Cummings, R.G., D.S. Brookshire, and W.D. Schulze, eds. 1986. Valuing Environmental Goods: An Assessment of the Contingent Valuation Method. Totowa, N.J.: Rowman and Allanheld.

Franklin, J.F. 1993. Preserving biodiversity: Species, ecosystems, or landscapes? Ecol. Appl. 3:202-205.

Gregory, R., S. Lichtenstein, and P. Slovic. 1993. Valuing environmental resources: A constructive approach. J. Risk Uncertainty 7:177-197.

Hageman, R. 1985. Valuing Marine Mammal Populations: Benefit Valuations in a Multi-Species Ecosystem. Admin. Rep. J-85-22. National Marine Fisheries Service, Southwest Fisheries Center, La Jolla, Calif.

Hagen, D., J. Vincent, and P. Welle. 1991. Benefits of Preserving Old-Growth Forests and the Northern Spotted Owl. Working paper prepared at Western Washington University, Bellingham, Wash.

Irwin, L.L., and T.B. Wigley. 1993. Toward an experimental basis for protecting forest wildlife. Ecol. Appl. 3:213-217.

Kareiva, P.M., J.G. Kingsolver, and R.B. Huey, eds. 1993. Biotic Interactions and Global Change. Sunderland, Mass.: Sinauer Associates.

LaRoe, E.T., III. 1993. Implementation of an ecosystem approach to endangered species conservation. Endangered Species Update 10 (3&4):3-6.

Mitchell, R.C., and R.T. Carson. 1988. Fusing Surveys to Value Public Goods: The Contingent Valuation Method. Resources for the Future, Washington, D.C.

Naiman, R.J., H. Décamps, and M. Pollock. 1993. The role of riparian corridors in maintaining regional biodiversity. Ecol. Appl. 3:209-212.

NOAA (U.S. National Oceanic and Atmospheric Administration) Panel on Contingent Evaluation. 1993. Fed. Reg. 58(10):4602-4614.

Norgaard, R.B. 1988. The rise of the global exchange economy and the loss of biological diversity. Pp. 206-211 in Biodiversity, E.O. Wilson, ed. Washington, D.C.: National Academy Press.

Norton, B. 1988. Commodity, amenity, and morality: The limits of quantification in biodiversity. Pp. 200-205 in Biodiversity, E.O. Wilson, ed. Washington, D.C.: National Academy Press.

NRC (National Research Council). 1992. Restoration of Aquatic Ecosystems: Science, Technology, and Public Policy. Washington, D.C.: National Academy Press.

NRC (National Research Council). 1993a. A Biological Survey for the Nation. Washington, D.C.: National Academy Press.

NRC (National Research Council). 1993b. Setting Priorities for Land Conservation. Washington, D.C.: National Academy Press.

Quigley, T.M., and S.E. McDonald. 1993. Ecosystem management in the Forest Service: Linkage to endangered species management. Endangered Species Update 10 (3&4):30-33.

Randall, A. 1988. What mainstream economists have to say about the value of biodiversity. Pp. 217-223 in Biodiversity, E.O. Wilson, ed. Washington, D.C.: National Academy Press.

Schulze, W.D., R.C. d'Arge, and D.S. Brookshire. 1981. Valuing environmental commodities: Some recent experiments. Land Econ. 57(2):151-172.

Stoll, J.R., and L.A. Johnson. 1984. Concepts of value, nonmarket valuation, and the case of the whooping crane. Trans. North Am. Wildl. Nat. Resour. Conf. 49:382-303.

Tversky, A., S. Sattah, and P. Slovic. 1988. Contingent weighting in judgement and choice. Psychol. Rev. 95:371-384.

Wilcove, D. 1993. Getting ahead of the extinction curve. Ecol. Appl. 3:218-220.

Wilson, E.O., ed. 1988. Biodiversity. Washington, D.C.: National Academy Press.

Wilson, E.O. 1992. The Diversity of Life. Cambridge, Mass.: Harvard University Press.

10

Beyond the Endangered Species Act

This chapter explores the ability of the Endangered Species Act to fulfill its purpose and how it might be complemented by other activities. In its drafting of the act, Congress clearly recognized the importance of habitats and ecosystems, because species—endangered and otherwise—cannot survive without them. To protect those habitats and ecosystems will require more than the ESA alone can provide, and that is also a topic of this chapter. The act's protections have helped listed species in many cases, although some species continue to become extinct. Ecosystem-level planning and management offer promise for addressing the conservation needs of a wider array of species than the traditional species-oriented approach and should be viewed as valuable complements to the ESA.

IS THE ESA WORKING?

Is the Endangered Species Act working? Critics of the act and its defenders debate this question as reauthorization looms. Not surprisingly, the answers differ depending on how the poser is affected by the act's provisions. To answer the question, we need to return to the objectives of the ESA as they are described in the act. Section 2(b) of the ESA states that the purposes of the act are "to provide a means whereby the ecosystems upon which endangered species and threatened species depend may be conserved, to provide a program for the conservation of such endangered species and threatened species, and to take such steps as may be appropriate to achieve the purposes of the treaties and conventions set forth in subsection (a) of this section." The essential questions are

- Do the ESA's protections reduce the likelihood of species extinction?
- Has the ESA successfully promoted species recovery?
- Are the ecosystems upon which threatened and endangered species depend being conserved?

REDUCING EXTINCTION

The difficulty with conclusively establishing extinction is manifest in several ways. First is the inherent problem of proving negative facts. No matter how many times species searches find nothing, there is always a chance that the object might appear. The black-footed ferret is a prime example. Hundreds of surveys were conducted throughout its historic range before it reappeared in Meeteetse, Wyoming, several years after the last known field population in South Dakota faded from view.

Second, information on an organism's status might be scanty because little effort has been taken to find and study it. In addition, finding rare species often requires specialized survey techniques that are not systematically applied, or the methods used might simply be inappropriate if little is known about the species. Finally, we might not have long enough time-series data or be looking in the right places (Taylor, 1993)—witness the proliferation of several threatened or endangered annual plants following heavy rains that broke the recent several-year drought in California's San Joaquin Valley. Hoover's woolly-star (*Eriastrum hooveri*) has been found in so many locations where it was formerly unknown or thought to be extirpated that resource agencies might soon submit a delisting petition (L. Saslaw, U.S. Bureau of Land Management, pers. commun.).

Although not conclusive, the comparison of rates of extinction between listed and candidate species, especially when remedial actions are clearly identifiable and feasible, indicates that the ESA helps to retard extinctions. Recovery actions encouraged by the ESA and supported by agency funding have helped to rescue several species from precarious status. A recent fact-sheet furnished by the Fish and Wildlife Service (FWS) Office of Endangered Species provides 30 ESA success stories. Several of them appear on the list of top 20 U.S. threatened and endangered animal species and top 20 plant species in order of federal and state expenditures for recovery in 1989 (CEQ, 1990), including several high-visibility symbols of the ESA's appeal, such as the bald eagle, grizzly bear, American peregrine falcon, whooping crane, southern sea otter, black-footed ferret, and California condor. Although FWS does not yet consider those species to be recovered, their chances for long-term survival have improved greatly. Others on the list that received substantial funding, such as the Puerto Rican parrot, would have fared much worse if not for strong intervention (Snyder et al., 1987), the recent damage caused by hurricane Andrew notwithstanding. Also, a recent GAO report (GAO, 1994) concludes that the National Wildlife Refuge system is contributing to the recovery of endangered species.

Combined with the mandates for federal agencies to avoid jeopardy and the ban on taking listed species, the affirmative steps provided in the ESA are helping to ease the risk to other species too. The status of the Utah prairie dog, piping plover, Oregon silver-spot butterfly, Aleutian Canada goose, Gila trout, greenback cutthroat trout, least Bell's vireo, California least tern, Virginia big-eared bat, red wolf, small whorled pagonia, and several others have improved greatly from the time of listing (FWS, 1990), although they are not out of danger. Others, such as the Florida panther, are still struggling despite intensive efforts to stabilize the remaining population. Without the protections and recovery actions required by the ESA, there is a strong, but hard-to-prove, possibility that most, if not all, of these and many other species would be closer to extinction than they are today. It is not possible to evaluate from available information whether any candidate species have improved without actions compelled by the act.

RECOVERY SUCCESS

In December 1990, the FWS published its first report to Congress on the endangered and threatened species recovery program. That report was prepared to meet a requirement in the 1988 amendments to the ESA, which were intended to improve recovery programs. Several provisions were enacted in response to public perceptions that the recovery planning and implementation process was not working very well (Fitzgerald and Meese, 1986; Clark and Harvey, 1988; GAO, 1988; Culbert and Blair, 1989).

The amendments provided for public involvement in recovery planning through review and commenting opportunities on draft recovery plans. Every 2 years, FWS and the National Marine Fisheries Service (NMFS) are to report to Congress on the status of recovery planning and implementation efforts. The resource agencies must set up systems for monitoring the status of recovered and delisted species. States are encouraged to use Section 6 funds to monitor the status of candidate species, and recovery expenditures must be reported annually. Congress tried to make recovery plans more useful by requiring them to identify site-specific management actions to achieve recovery goals, to estimate the time needed for recovery assuming sufficient funds are available for implementation, to estimate costs required for successful implementation, and to set measurable recovery criteria that enable FWS and NMFS to evaluate recovery success.

According to FWS (1992), recovery is the process by which the decline of an endangered or threatened species is arrested or reversed, and threats to its survival are neutralized. The goal of the process is to achieve sufficient self-sustaining wild populations of listed species to ensure their survival in nature.

FWS aims to (1) identify ecosystems and organisms facing the greatest degree of threat, (2) determine steps necessary to reduce or eliminate the threats, and (3) apply the resources available to the highest priority recovery tasks. Restoration to the point where species can be delisted is the ultimate objective, although removal from the list is not a reasonable measure of short-term success (FWS, 1992). FWS asserts that a more realistic metric of recovery efforts is the number of species whose decline has been arrested and the population stabilized.

The recovery process starts with development of a recovery plan. The purpose is to develop species-specific recovery goals; identify needed biological information, including the status of the species; and set forth management tasks required to recover the species. FWS maintains that coordination among federal, state, and local agencies; academic researchers; conservation organizations; private individuals; and major land users might be the most essential ingredient for the development and implementation of an effective recovery program. The agency further states that it emphasizes cooperation and teamwork among all involved parties.

As described in Chapter 4, the resource agencies had approved 411 recovery plans covering 513 species as of March 1993—54% of the 956 U.S. species listed at that time. The percentage of species having approved recovery plans has dropped somewhat from 1990, when 352 out of 581 listed species (61%) were covered (FWS, 1990). This is probably attributable to recent accelerated listing actions and agency preoccupation with several manpower-intensive listing and recovery efforts (J. Bartell, FWS, pers. commun.). Chapter 4 also describes FWS's recovery backlog, which is likely to expand in future years as the pace of listings increases in response to the recent settlement over Category 1 candidates and "warranted but precluded" species.

Because most recovery plans were prepared before the 1988 amendments were passed, we only recently began to see the effects of the new requirements. In April 1993, 80 species had revised plans that were intended to comply with the 1988 amendments; another 63 species had draft first plans or draft revised plans.

FWS broke out the listed species covered in its 1990 (581 species) and 1992[1] (711 species) reports to Congress into the following groups:

---

[1]The latest information available to the committee.

	1990	**1992**
Improving	57 (9.8%)	69 (9.7%)
Stable	181 (31.2%)	201 (28.3%)
Declining	219 (37.7%)	232 (32.6%)
Extinct	11 (1.9%)	14 (2.0%)
Unknown	113 (19.4%)	195 (27.4%)

These results are interpreted differently by various observers even within the environmental community. Some are encouraged (Irvin, 1993) by the number of species stabilized or improving, and others lament the lack of recovery plans for many species and the high numbers known to be declining or whose status is unknown. Critics of the ESA point to the low numbers of delistings due to successful recovery and the high costs required to get there for many others (Desiderio, 1993). Only six species have been delisted due to successful recovery: the Palau dove, Palau owl, Palau flycatcher, the Rydberg milk-vetch, the Atlantic coast population of the brown pelican, and the gray whale. Several recoveries (or improvements in status) happened because more individuals were found after the species were listed. For example, the upgrading of the formerly endangered snail darter of Tennessee (a fish, *Percina tanasi*) was largely due to the discovery of additional populations in different rivers from where it was originally discovered (Etnier and Starnes, 1993). In other cases, progress toward recoveries might have been due at least in part to actions unrelated to the ESA, such as the banning of DDT.

Expenditures for recovery have increased in recent years, although they are far less than needed to attain recovery goals for the bulk of listed species (DOI, 1990; Jackson, 1992). FWS's budget for recovery programs was $10.4 million in FY 1990 and $39.7 million in FY 1995, although additional funds were available from other state and federal agencies. FWS estimates that approximately $177 million was spent for endangered-species conservation programs in 1991 (FWS, 1992), but other funding approaches or even exceeds that figure: for example, the Bonneville Power Administration (BPA) estimated that its 1994 expenditures on salmon conservation (including some stocks listed as endangered under the ESA) in the Columbia River basin amounted to about $350 million, of which about $300 million was due to lost power revenues (NRC, 1995). The importance of these numbers is not easy to interpret: the BPA expenditure was 1% of its 1994 revenues. That portion of the U.S. budget would be about $10 billion per year.

Another concern is that what FWS means by stable, improving, or declining is unclear. This lack of precision mirrors reservations expressed by Rohlf (1991) about the absence of clearly stated thresholds to define differences among threatened, endangered, and secure (recovered) species. Although these categories are qualitative designations, they are not very useful in comparing species trends, because no measurable criteria are used to define them, as discussed in Chapter 8. For example, what percentage change in population size over what period constitutes a decline or improvement? Within what range of variation around a midpoint must a population remain to be considered stable? For how long? If the population size remains constant for several years but it is above or below the numbers when it was listed, is this considered stable or does it constitute a trend in either direction? It is not apparent that calling a species stable allows someone to distinguish between species that are still at critically low levels and those that are more abundant and less at risk. Chapter 8 offers guidance and encouragement to FWS to assign measurable criteria to these groupings in future reports.

The committee concludes that although the ESA has undoubtedly protected some species from extinction, the overall effectiveness of recovery plans cannot be quantified at this time. In part, this is because recovery plans can take a long time to work or to fail, especially where long-lived species are concerned. In addition, as mentioned elsewhere, many recovery plans are implemented only after long delays, and often not at all.

PROTECTION OF ECOSYSTEMS

Although it is clear that the ESA has resulted in the protection of some ecosystems on which endangered species depend, our ability to quantify the contribution FWS and NMFS regulatory programs make in protecting them is limited. Lacking effective metrics, we are left to compare the relative advantages and disadvantages of species-by-species management so common in the listing and recovery programs with more regional ecosystem-oriented efforts. Although the purpose of the act is in part to protect ecosystems on which threatened and endangered species depend, public support and congressional appropriations are more clearly linked with the need to protect individual species having broad social or cultural significance (Rohlf, 1991). Traditional approaches sometimes have been necessitated by the urgent need to save certain species dangerously close to extinction during the early years of the act's tenure.

The historic focus of FWS and NMFS regulatory programs on saving individual species is acknowledged by the agencies (FWS, 1990). Listings have occurred mostly on a species-by-species schedule with some exceptions, including plants from rare communities in Hawaii, Florida, and Puerto Rico. Recent litigation settlement agreements committing FWS to act on several hundred candidate species provide that FWS will use a multispecies, ecosystem-based approach for listing proposals and critical habitat designation when biologically appropriate. Also, habitat conservation planning, although seldom used in practice, has the potential to be effective in protecting ecosystems and has realized that potential in a few cases (Chapter 4).

With recovery protocols for 513 species contained in 411 plans, about one quarter of approved plans address multiple species. Without reviewing each plan, the committee could not tell how many of those prepared for more than one species try to meet common ecosystem objectives. Examples of recovery plans that feature an ecosystem perspective include Ash Meadows (five animals and seven plants), Maui-Molokai forest birds (seven species), and California Channel Island species (FWS, 1990). Plan development is under way for six plants and four animals in the San Joaquin Valley of California, where collaborators are trying to take a landscape view covering several biological community types and a multitude of competing land uses. On February 23, 1994, FWS released a draft recovery plan for aquatic species in the Snake River. The draft plan included 16 fish taxa and 42 mollusks (including five listed snails).

Because listings and recovery plans concentrate on individual species does not mean that the ecosystems where they play constituent roles are not being protected. Indeed, if it did not protect ecosystems, at least to some degree, the ESA would be less controversial than it is, because that protection includes a prohibition of some economic activities. Perhaps the best-known example is the limitations on forest harvest to protect the northern spotted owl's ecosystem. Even where the ESA has led to incomplete or even no protection of an ecosystem, it has focused attention on the nature and biological significance of many ecosystems.

Unless the populations are being artificially maintained, recovery success depends on proper consideration of how the species interacts with surrounding biotic and physical environmental factors. Self-sustaining wild populations require sufficient natural habitat, including food, water, shelter, movement corridors, and the many other features essential for survival and reproduction (Morrison et al., 1992). As discussed above, about 4% of listed species have already recovered or are close enough to downlisting or recovery thresholds for FWS to take administrative action. By definition, if the species are recovered, enough ecosystem-level values must be present to ensure long-term persistence. The question is whether a broader, more systematic look at the recovery needs of multiple species would improve the performance of FWS's recovery program.

In addition to programs under the ESA, other federal agencies have land-management authorities and responsibilities, and their activities—some under congressional mandates—include land acquisition, wildlife management, management of parks and wilderness areas, protection of wetlands, and prevention of environmental pollution. These activities also result in protection of ecosystems; the responsible agencies include the National Park

Service, the Bureau of Land Management, the Army Corps of Engineers, the Forest Service, the Natural Resources Conservation Service (formerly the Soil Conservation Service), the Environmental Protection Agency, and the Bureau of Reclamation.

THE FUTURE: BEYOND THE ENDANGERED SPECIES ACT

The ESA and other existing programs will not *by themselves* prevent all future extinctions of species in the United States. It appears to the committee that Congress intended the ESA to be a safety net to protect endangered species, and the committee concludes that that is its proper role. However—and this is not entirely the fault of the ESA—species often will be in serious trouble by the time they receive ESA protection. The director of the Fish and Widlife Service, Mollie Beattie, holds a similar view. She said (as quoted by the *Jackson Hole News*, February 8, 1995): "The Endangered Species Act is . . . a law that plays in when local planning and zoning, state fish and wildlife efforts, the Clean Water Act, and Clean Air Act haven't worked. It is the emergency room of conservation policy." If species extinctions are to be prevented, a broader management approach will be needed to complement the ESA's protections. A few thoughts on that broader approach—ecosystem management—are appropriate here.

The goal of an ecosystem-based approach to managing natural resources is to maintain biological diversity by recognizing the value of protecting an array of biological communities and habitat types within a larger landscape context (Hunter, 1990). Ecosystem-focused programs are probably most useful when individual elements of biological communities are not in so much trouble that they need narrowly targeted management efforts.

Using an ecosystem perspective[2] for endangered-species-conservation planning offers several advantages. First, species needs are viewed in the context of surrounding land uses, rather than within the limits of their currently occupied habitat. Because surrounding land uses and the distribution of habitat patches among them can strongly influence species welfare (Hunter, 1990), resource managers can identify future opportunities and constraints.

Second, the complexity of the problems facing managers who develop and implement strategies for conservation of endangered species requires new concepts (LaRoe, 1993). The expected rapid pace of new listings coupled with funding limitations places a premium on approaches that address the needs of different species simultaneously. Recently developed tools, such as advances in remote sensing, population-viability-analysis models, decision-analysis methods (see Chapter 8), and geographic information sys-

---

[2]See discussion of ecosystem management in Chapter 9.

tems, increase our ability to analyze complex problems involving interactions among diverse influences and the implications of various possible solutions. Gap analysis (Scott et al., 1987) can provide a process to identify areas of high biological diversity or those containing several protected species with overlapping ranges. It can also be used to develop a coarse overview of resource status and land uses across political boundaries. Additional management approaches related to ecosystem management show promise. They include the following:

• *Reconstruction or rehabilitation of ecosystems.* Restoration ecology is a growing discipline. Many ecosystem functions have been improved or restored by such activities, and reconstruction or rehabilitation of ecosystem functioning holds much promise for the protection of endangered species. It is not usually possible to return an ecosystem to some prior pristine condition, however. Many ecosystems have been so altered that it is difficult to decide what prior condition we might want to return to. The trajectory taken by the ecosystem to get to its current condition is not retraceable in the way that a highway is, because many events occur in an ecosystem's history that are not precisely reversible. Genetic variability is lost; evolution occurs; exotic species are introduced; human populations in the region increase, and people develop dependence on a variety of modern technologies, cultures, and economic systems; and other natural and anthropogenic environmental changes affect the range of biophysical and socioeconomic possibilities for future states of the system. In brief, the past provides opportunities for the future but also constrains it. Thus, attempts to rehabilitate ecosystem functioning should keep these constraints in mind, so that inappropriately high expectations are not generated.

• *Mixed management plans.* Often, resource managers manage areas either for protection of biota or for human use. It is increasingly difficult to keep people and the effects of their activities separate from wildlife sanctuaries. Although such sanctuaries (e.g., national parks, wilderness areas, wildlife refuges, marine sanctuaries) are indispensable for protecting endangered species, greater attention needs to be paid to developing mixed-use areas. Those would be urban recreation areas or residential and commercial developments adjacent to untrammeled areas designed to improve opportunities for wildlife while maintaining opportunities for human activities. Although the value of this approach is becoming increasingly recognized, its development is still in the early stages.

• *Cooperative management.* Various experiences with cooperative management—the sharing of planning and decision making by various government and nongovernment groups—have had some success. To some degree, habitat conservation plans represent an example of this approach, but it is likely that cooperative management will be necessary in cases where the strict requirements of the Endangered Species Act have not yet been applied. It is important to include the major interested parties without having so many interests involved that consensus is difficult to reach.

• *Revised economic accounting.* Too often, economic calculations underlying public and private decision making are incomplete. Often, they cover too short a time span, and they often exclude nonmarket values. A short-term loss might turn into a long-term gain: for example, losing an economic activity today might provide opportunities for greater economic activities of different types at some time in the future. Again, the validity of expanding economic accounting to cover longer periods and to include nonmarket values is becoming more widely recognized, but it is still in the early stages of development.

SCIENCE, POLICY, AND THE ESA

This committee was asked to review the scientific aspects of the ESA, and it has done so. It has not uncovered any major scientific issue that seriously hinders the implementation of the act, although its review has suggested several scientific improvements. Many of the conflicts and disagreements about the ESA do not appear to be based on scientific issues. Instead, they appear to result because the act—in the committee's opinion designed as a safety net or act of last resort—is called into play when other policies and management strategies or their failures, or human activities in general, have led to the endangerment of species and populations. In some cases, policies and programs have been based on sound science, but other factors have prevented them from working. The committee does not see any likelihood that those endangerments will soon cease to occur or that the ESA can or should be expected to prevent them from occurring. It therefore concludes that any coherent, successful program to prevent species endangerments and to protect the nation's biological diversity is going to require more enlightened commitments on the part of all major parties to achieve success.

To conserve natural habitats, approaches must be developed that rely on cooperation and innovative procedures; examples provided by the ESA are habitat conservation plans and natural community conservation planning. But those are only a beginning. Many other approaches have been discussed in various forums. They include cooperative management (sharing decision-making authority among several governmental and nongovernmental groups), transfer of development credits, mitigation banks, tax incentives, and conservation easements.

An analysis of these and other policy and management options is beyond this committee's charge, but sound science alone will not lead to successful prevention of many species extinctions, conservation of biological diversity, and reduced economic and social uncertainty and disruption. But sound science is an essential starting point. Combined with innovative and workable policies, it can help to solve these and related problems.

REFERENCES

CEQ (Council on Environmental Quality). 1990. Twenty-first Annual Report. Washington, D.C.: U.S. Government Printing Office.

Clark, T.W., and A.H. Harvey. 1988. Implementing endangered species recovery policy: Learning as we go. Endangered Species Update 5:35-42.

Culbert, R., and R.B. Blair. 1989. Recovery planning and endangered species. Endangered Species Update 6:2-8.

Desiderio, M. 1993. The ESA: Facing hard truths and advocating responsible reform. Nat. Resour. Environ. 8:37, 41-42.

DOI (U.S. Department of the Interior) Office of the Inspector General. 1990. Audit Report: The Endangered Species Program, U.S. Fish and Wildlife Service, Washington, D.C.

Etnier, D.A., and W.C. Starnes. 1993. The Fishes of Tennessee. Knoxville, Tenn.: University of Tennessee Press.

Fitzgerald, J., and G.M. Meese. 1986. Saving Endangered Species, Amending and Implementing the Endangered Species Act. Defenders of Wildlife, Washington, D.C. 36 pp.

FWS (U.S. Fish and Wildlife Service). 1990. Report to Congress, Endangered and Threatened Species Recovery Program. Washington, D.C.: U.S. Government Printing Office.

FWS (U.S. Fish and Wildlife Service). 1992. Report to Congress, Endangered and Threatened Species Recovery Program. Washington, D.C.: U. S. Government Printing Office.

GAO (General Accounting Office). 1988. Endangered Species Management Improvements Could Enhance Recovery Programs. U.S. General Accounting Office, Washington, D.C.

GAO (General Accounting Office). 1994. National Wildlife Refuge System: Contributions Being Made to Endangered Species Recovery. U.S. General Accounting Office, Washington, D.C.

Hunter, M.L. 1990. Wildlife, Forests, and Forestry, Principles of Managing Forests for Biological Diversity. Englewood Cliffs, N.J.: Regents/Prentice Hall. 370 pp.

Irvin, W.R. 1993. The Endangered Species Act: Keeping Every Cog and Wheel. Nat. Resour. Environ. 8:36, 38-40, 76.

Jackson, T.C. 1992. All creatures great and small. Legal Times (Dec. 7):20-23.

LaRoe, E.T., III. 1993. Implementation of an Ecosystem Approach to Endangered Species Conservation. Endangered Species Update 10:3-6.

Morrison, M.L., B.G. Marcot, and R.W. Mannan. 1992. Wildlife-Habitat Relationships, Concepts and Applications. Madison, Wisc.: University of Wisconsin Press. 343 pp.

NRC (National Research Council). 1995. Upstream: Salmon and Society in the Pacific Northwest. Washington, D.C.: National Academy Press

Rohlf, D.J. 1991. Six biological reasons why the Endangered Species Act doesn't work—and what to do about it. Conserv. Biol. 5:273-282.

Scott, J.M., B. Csuti, J.D. Jacobi, and J.E. Estes. 1987. Species richness. Bioscience 39:782-788.

Snyder, N.F.R., J.W. Wiley, and C.B. Kepler. 1987. The Parrots of Luquillo: Natural History and Conservation of the Puerto Rican Parrot. Western Foundation of Vertebrate Zoology, Los Angeles, Calif. 384 pp.

Taylor, R.J. 1993. Biological Uncertainty in the Endangered Species Act. Nat. Resour. Environ. 8:6-9, 58-59.

Taylor, S. 1993. Practical ecosystem management for plants and animals. Endangered Species Update 10:26-29.

APPENDIXES

APPENDIX
A

Congress of the United States
Washington, DC 20515

November 27, 1991

Dr. Frank Press, Director
National Academy of Sciences
2101 Constitution Avenue, N.W.
Washington, D.C. 20418

Dear Doctor Press:

We are writing to request the National Academy of Sciences to
conduct a study of several issues related to the Endangered
Species Act (ESA).

Almost two decades have passed since the Endangered Species Act
was first enacted by Congress. During its implementation,
several important and highly complex biological issues have
arisen. We propose that the Academy review the following issues
and evaluate how they relate to the overall purposes of the
Endangered Species Act.

1. DEFINITION OF SPECIES. One of the fundamental conceptual
building blocks of the ESA is the concept of "species". There
have been considerable discussions within scientific circles on
how to identify the appropriate taxonomic units to achieve the
purposes of the Act, with particular focus on the question of
populations and population segments. We would therefore
appreciate a review of the manner in which the term "species" has
been implemented in order to evaluate how to identify those units
that will best serve the purposes of the Act.

2. CONFLICTS BETWEEN SPECIES. Proposals to list several species
of salmon in the Pacific northwest have generated difficult
issues about how to reconcile the conservation needs of different

November 27, 1991
Page Two

listed species where those needs may conflict. Closely related
questions have also arisen about how to account for the effects
of conservation efforts under the ESA on other potentially
imperiled species and other environmental values generally. We
would welcome the recommendations of the Academy on the severity
of these types of problems and how best to address them.

3. ROLE OF HABITAT CONSERVATION. Many observers debate the role
that habitat protection could play in achieving the purposes of
the Act in relation to the current emphasis on protecting
individual species. The Act currently provides several
mechanisms for the protection of habitat through the designation
of critical habitat, the adoption of habitat conservation plans
and other provisions. We would appreciate an evaluation of the
role of habitat protection in contributing to the conservation of
species, both wide ranging and narrowly distributed, and a review
of the relationship of habitat protection mechanisms of the Act
to its other requirements.

4. RECOVERY PLANNING. Another issue closely related to item
three is whether recovery planning is properly integrated with
the other requirements of the ESA. For example, some
commentators have suggested that the designation of critical
habitat before the development of recovery plans may not be
prudent and that the two efforts ought to be more closely
integrated. We would therefore appreciate a review of the role
of recovery planning in the Act and any recommendations on how
recovery planning could better contribute to the purposes of the
Act.

5. RISK. Judgments about acceptable risk pervade many of the
decisions required by the Endangered Species Act, including those
relating to whether and how to list a species and what
constitutes jeopardy, adverse modification, reasonable and
prudent alternatives, taking, conservation and recovery. We
would appreciate a review of the role that risk has played in
decisionmaking under the ESA. We would also appreciate a review
of whether different levels of risk ought to apply to different
types of decisions (and the practical methods that might be
employed to assess risk) to better achieve the purposes of the
Act while providing flexibility in appropriate circumstances to
accommodate other objectives as well.

6. ISSUES OF TIMING. One final question that deserves review
relates to the timing of certain key decisions under the Act. We
would welcome the recommendations of the Academy on how to
improve the timing of decisions under the Act in order to better
serve its purposes while minimizing otherwise unintended
consequences.

November 27, 1991
Page Three

We appreciate your attention to this request and believe that a
report by the National Academy could make an important
contribution to the public discourse on the Endangered Species
Act.   Should you have any questions about it, please do not
hesitate to contact Will Stelle, Sandy Mathiesen or Mark Walker
of our staffs.

With kind regards.

                              Sincerely,

Thomas Foley   Mark O. Hatfield
  The Speaker   U.S. Senator

        Gerry. E. Studds, Chairman
         Subcommittee on Fisheries and Wildlife
         Conservation and the Environment

APPENDIX
B

ENDANGERED SPECIES ACT OF 1973

As Amended through the

100th Congress

U.S. Fish and Wildlife Service
U.S. Department of the Interior
Washington, D.C. 20240
1988

TABLE OF CONTENTS

ENDANGERED SPECIES ACT OF 1973*

FINDINGS,₌PURPOSES, AND POLICY

SEC. 2. (a) FINDINGS.—The Congress finds and declares that—

(1) various species of fish, wildlife, and plants in the United States have been rendered extinct as a consequence of economic growth and development untempered by adequate concern and conservation;

(2) other species of fish, wildlife, and plants have been so depleted in numbers that they are in danger of or threatened with extinction;

(3) these species of fish, wildlife, and plants are of esthetic, ecological, educational, historical, recreational, and scientific value to the Nation and its people;

(4) the United States has pledged itself as a sovereign state in the international community to conserve to the extent practicable the various species of fish or wildlife and plants facing extinction, pursuant to—

(A) migratory bird treaties with Canada and Mexico;

(B) the Migratory and Endangered Bird Treaty with Japan;

(C) the Convention on Nature Protection and Wildlife Preservation in the Western Hemisphere;

(D) the International Convention for the Northwest Atlantic Fisheries;

(E) the International Convention for the High Seas Fisheries of the North Pacific Ocean;

(F) the Convention on International Trade in Endangered Species of Wild Fauna and Flora; and

(G) other international agreements; and

(5) encouraging the States and other interested parties, through Federal financial assistance and a system of incentives, to develop and maintain conservation programs which meet national and international standards is a key to meeting the Nation's international commitments and to better safeguarding, for the benefit of all citizens, the Nation's heritage in fish, wildlife, and plants.

(b) PURPOSES.—The purposes of this Act are to provide a means whereby the ecosystems upon which endangered species and threatened species depend may be conserved, to provide a program for the conservation of such endangered species and threatened species, and to take such steps as may be appropriate to achieve

---

*As amended by P.L. 94–325, June 30, 1976; P.L. 94–359, July 12, 1976; P.L. 95–212, December 19, 1977; P.L. 95–632, November 10, 1978; P.L. 96–159, December 28, 1979; P.L. 97–304, October 13, 1982; P.L. 98–327, June 25, 1984; and P.L. 100–478, October 7, 1988.

the purposes of the treaties and conventions set forth in subsection (a) of this section.

(c) POLICY.—(1) It is further declared to be the policy of Congress that all Federal departments and agencies shall seek to conserve endangered species and threatened species and shall utilize their authorities in furtherance of the purposes of this Act.

(2) It is further declared to be the policy of Congress that Federal agencies shall cooperate with State and local agencies to resolve water resource issues in concert with conservation of endangered species.

DEFINITIONS

SEC. 3. For the purposes of this Act—

(1) The term "alternative courses of action" means all alternatives and thus is not limited to original project objectives and agency jurisdiction.

(2) The term "commercial activity" means all activities of industry and trade, including, but not limited to, the buying or selling of commodities and activities conducted for the purpose of facilitating such buying and selling: *Provided, however,* That it does not include exhibitions of commodities by museums or similar cultural or historical organizations.

(3) The terms "conserve," "conserving," and "conservation" mean to use and the use of all methods and procedures which are necessary to bring any endangered species or threatened species to the point at which the measures provided pursuant to this Act are no longer necessary. Such methods and procedures include, but are not limited to, all activities associated with scientific resources management such as research, census, law enforcement, habitat acquisition and maintenance, propagation, live trapping, and transplantation, and, in the extraordinary case where population pressures within a given ecosystem cannot be otherwise relieved, may include regulated taking.

(4) The term "Convention" means the Convention on International Trade in Endangered Species of Wild Fauna and Flora, signed on March 3, 1973, and the appendices thereto.

(5)(A) The term "critical habitat" for a threatened or endangered species means—

(i) the specific areas within the geographical area occupied by the species, at the time it is listed in accordance with the provisions of section 4 of this Act, on which are found those physical or biological features (I) essential to the conservation of the species and (II) which may require special management considerations or protection; and

(ii) specific areas outside the geographical area occupied by the species at the time it is listed in accordance with the provisions of section 4 of this Act, upon a determination by the Secretary that such areas are essential for the conservation of the species.

(B) Critical habitat may be established for those species now listed as threatened or endangered species for which no critical habitat has heretofore been established as set forth in subparagraph (A) of this paragraph.

(C) Except in those circumstances determined by the Secretary, critical habitat shall not include the entire geographical area which can be occupied by the threatened or endangered species.

(6) The term "endangered species" means any species which is in danger of extinction throughout all or a significant portion of its range other than a species of the Class Insecta determined by the Secretary to constitute a pest whose protection under the provisions of this Act would present an overwhelming and overriding risk to man.

(7) The term "Federal agency" means any department, agency, or instrumentality of the United States.

(8) The term "fish or wildlife" means any member of the animal kingdom, including without limitation any mammal, fish, bird (including any migratory, nonmigratory, or endangered bird for which protection is also afforded by treaty or other international agreement), amphibian, reptile, mollusk, crustacean, arthropod or other invertebrate, and includes any part, product, egg, or offspring thereof, or the dead body or parts thereof.

(9) The term "foreign commerce" includes, among other things, any transaction—

  (A) between persons within one foreign country;

  (B) between persons in two or more foreign countries;

  (C) between a person within the United States and a person in a foreign country; or

  (D) between persons within the United States, where the fish and wildlife in question are moving in any country or countries outside the United States.

(10) The term "import" means to land on, bring into, or introduce into, or attempt to land on, bring into, or introduce into, any place subject to the jurisdiction of the United States, whether or not such landing, bringing, or introduction constitutes an importation within the meaning of the customs laws of the United States.

(11) The term "permit or license applicant" means, when used with respect to an action of a Federal agency for which exemption is sought under section 7, any person whose application to such agency for a permit or license has been denied primarily because of the application of section 7(a) to such agency action.

(12) "The term person means an individual, corporation, partnership, trust, association, or any other private entity; or any officer, employee, agent, department, or instrumentality of the Federal Government, of any State, municipality, or political subdivision of a State, or of any foreign government; any State, muncipality, or political subdivision of a State; or any other entity subject to the jurisdiction of the United States."

(13) The term "plant" means any member of the plant kingdom, including seeds, roots and other parts thereof.

(14) The term "Secretary" means, except as otherwise herein provided, the Secretary of the Interior or the Secretary of Commerce as program responsibilities are vested pursuant to the provisions of Reorganization Plan Numbered 4 of 1970; except that with respect to the enforcement of the provisions of this Act and the Convention which pertain to the importation or exportation of terrestrial plants, the term also means the Secretary of Agriculture.

(15) The term "species" includes any subspecies of fish or wildlife or plants, and any distinct population segment of any species or vertebrate fish or wildlife which interbreeds when mature.

(16) The term "State" means any of the several States, the District of Columbia, the Commonwealth of Puerto Rico, American Samoa, the Virgin Islands, Guam, and the Trust Territory of the Pacific Islands.

(17) The term "State agency" means any State agency, department, board, commission, or other governmental entity which is responsible for the management and conservation of fish, plant, or wildlife resources within a State.

(18) The term "take" means to harass, harm, pursue, hunt, shoot, wound, kill, trap, capture, or collect, or to attempt to engage in any such conduct.

(19) The term "threatened species" means any species which is likely to become an endangered species within the foreseeable future throughout all or a significant portion of its range.

(20) The term "United States," when used in a geographical context, includes all States.

DETERMINATION OF ENDANGERED SPECIES AND THREATENED SPECIES

SEC. 4. (a) GENERAL.—(1) The Secretary shall by regulation promulgated in accordance with subsection (b) determine whether any species is an endangered species or a threatened species because of any of the following factors:

(A) the present or threatened destruction, modification, or curtailment of its habitat or range;

(B) overutilization for commercial, recreational, scientific, or educational purposes;

(C) disease or predation;

(D) the inadequacy of existing regulatory mechanisms;

(E) other natural or manmade factors affecting its continued existence.

(2) With respect to any species over which program responsibilities have been vested in the Secretary of Commerce pursuant to Reorganization Plan Numbered 4 of 1970—

(A) in any case in which the Secretary of Commerce determines that such species should—

(i) be listed as an endangered species or a threatened species, or

(ii) be changed in status from a threatened species to an endangered species, he shall so inform the Secretary of the Interior, who shall list such species in accordance with this section;

(B) in any case in which the Secretary of Commerce determines that such species should—

(i) be removed from any list published pursuant to subsection (c) of this section, or

(ii) be changed in status from an endangered species to a threatened species, he shall recommend such action to the Secretary of the Interior, and the Secretary of the Interior, if he concurs in the recommendation, shall implement such action; and

(C) the Secretary of the Interior may not list or remove from any list any such species, and may not change the status of any such species which are listed, without a prior favorable determination made pursuant to this section by the Secretary of Commerce.

(3) The Secretary, by regulation promulgated in accordance with subsection (b) and to the maximum extent prudent and determinable—

(A) shall, concurrently with making a determination under paragraph (1) that a species is an endangered species or a threatened species, designate any habitat of such species which is then considered to be critical habitat; and

(B) may, from time-to-time thereafter as appropriate, revise such designation.

(b) BASIS FOR DETERMINATIONS.—(1)(A) The Secretary shall make determinations required by subsection (a)(1) solely on the basis of the best scientific and commercial data available to him after conducting a review of the status of the species and after taking into account those efforts, if any, being made by any State or foreign nation, or any political subdivision of a State or foreign nation, to protect such species, whether by predator control, protection of habitat and food supply, or other conservation practices, within any area under its jurisdiction, or on the high seas.

(B) In carrying out this section, the Secretary shall give consideration to species which have been—

(i) designated as requiring protection from unrestricted commerce by any foreign nation, or pursuant to any international agreement; or

(ii) identified as in danger of extinction, or likely to become so within the foreseeable future, by any State agency or by any agency of a foreign nation that is responsible for the conservation of fish or wildlife or plants.

(2) The Secretary shall designate critical habitat, and make revisions thereto, under subsection (a)(3) on the basis of the best scientific data available and after taking into consideration the economic impact, and any other relevant impact, of specifying any particular area as critical habitat. The Secretary may exclude any area from critical habitat if he determines that the benefits of such exclusion outweigh the benefits of specifying such area as part of the critical habitat, unless he determines, based on the best scientific and commercial data available, that the failure to designate such area as critical habitat will result in the extinction of the species concerned.

(3)(A) To the maximum extent practicable, within 90 days after receiving the petition of an interested person under section 553(e) of title 5, United States Code, to add a species to, or to remove a species from, either of the lists published under subsection (c), the Secretary shall make a finding as to whether the petition presents substantial scientific or commercial information indicating that the petitioned action may be warranted. If such a petition is found to present such information, the Secretary shall promptly commence a review of the status of the species concerned. The Secretary shall promptly publish each finding made under this subparagraph in the Federal Register.

(B) Within 12 months after receiving a petition that is found under subparagraph (A) to present substantial information indicating that the petitioned action may be warranted, the Secretary shall make one of the following findings:

(i) The petitioned action is not warranted, in which case the Secretary shall promptly publish such finding in the Federal Register.

(ii) The petitioned action is warranted in which case the Secretary shall promptly publish in the Federal Register a general notice and the complete text of a proposed regulation to implement such action in accordance with paragraph (5).

(iii) The petitioned action is warranted but that—

(I) the immediate proposal and timely promulgation of a final regulation implementing the petitioned action in accordance with paragraphs (5) and (6) is precluded by pending proposals to determine whether any species is an endangered species or a threatened species, and

(II) expeditious progress is being made to add qualified species to either of the lists published under subsection (c) and to remove from such lists species for which the protections of the Act are no longer necessary.

in which case the Secretary shall promptly publish such finding in the Federal Register, together with a description and evaluaton of the reasons and data on which the finding is based.

(C)(i) A petition with respect to which a finding is made under subparagraph (B)(iii) shall be treated as a petition that is resubmitted to the Secretary under subparagraph (A) on the date of such finding and that presents substantial scientific or commerical information that the petitioned action may be warranted.

(ii) Any negative finding described in subparagraph (A) and any finding described in subparagraph (B) (i) or (iii) shall be subject to judicial review.

(iii) The Secretary shall implement a system to monitor effectively the status of all species with respect to which a finding is made under subparagraph (B)(iii) and shall make prompt use of the authority under paragraph 7 to prevent a significant risk to the well being of any such species.

(D)(i) To the maximum extent practicable, within 90 days after receiving the petition of an interested person under section 553(e) of title 5, United States Code, to revise a critical habitat designation, the Secretary shall make a finding as to whether the petition presents substantial scientific information indicating that the revision may be warranted. The Secretary shall promptly publish such finding in the Federal Register.

(ii) Within 12 months after receiving a petition that is found under clause (i) to present substantial information indicating that the requested revision may be warranted, the Secretary shall determine how he intends to proceed with the requested revision, and shall promptly publish notice of such intention in the Federal Register.

(4) Except as provided in paragraphs (5) and (6) of this subsection, the provisions of section 553 of title 5, United States Code (relating

to rulemaking procedures), shall apply to any regulation promulgated to carry out the purposes of this Act.

(5) With respect to any regulation proposed by the Secretary to implement a determination, designation, or revision referred to in subsection (a) (1) or (3), the Secretary shall—

(A) not less than 90 days before the effective date of the regulation—

(i) publish a general notice and the complete text of the proposed regulation in the Federal Register, and

(ii) give actual notice of the proposed regulation (including the complete text of the regulation) to the State agency in each State in which the species is believed to occur, and to each county or equivalent jurisdiction in which the species is believed to occur, and invite the comment of such agency, and each such jurisdiction, thereon;

(B) insofar as practical, and in cooperation with the Secretary of State, give notice of the proposed regulation to each foreign nation in which the species is believed to occur or whose citizens harvest the species on the high seas, and invite the comment of such nation thereon;

(C) give notice of the proposed regulation to such professional scientific organizations as he deems appropriate;

(D) publish a summary of the proposed regulation in a newspaper of general circulation in each area of the United States in which the species is believed to occur; and

(E) promptly hold one public hearing on the proposed regulation if any person files a request for such a hearing within 45 days after the date of publication of general notice.

(6)(A) Within the one-year period beginning on the date on which general notice is published in accordance with paragraph (5)(A)(i) regarding a proposed regulation, the Secretary shall publish in the Federal Register—

(i) if a determination as to whether a species is an endangered species or a threatened species, or a revision of critical habitat, is involved, either—

(I) a final regulation to implement such determination,

(II) a final regulation to implement such revision or a finding that such revision should not be made,

(III) notice that such one-year period is being extended under subparagraph (B)(i), or

(IV) notice that the proposed regulation is being withdrawn under subparagraph (B)(ii), together with the finding on which such withdrawal is based; or

(ii) subject to subparagraph (C), if a designation of critical habitat is involved, either—

(I) a final regulation to implement such designation, or

(II) notice that such one-year period is being extended under such subparagraph.

(B)(i) If the Secretary finds with respect to a proposed regulation referred to in subparagraph (A)(i) that there is substantial disagreement regarding the sufficiency or accuracy of the available data relevant to the determination or revision concerned the Secretary may extend the one-year period specified in subparagraph (A) for not more than six months for purposes of soliciting additional data.

(ii) If a proposed regulation referred to in subparagraph (a)(i) is not promulgated as a final regulation within such one-year period (or longer period if extension under clause (i) applies) because the Secretary finds that there is not sufficient evidence to justify the action proposed by the regulation the Secretary shall immediately withdraw the regulation. The finding on which a withdrawal is based shall be subject to judicial review. The Secretary may not propose a regulation that has previously been withdrawn under this clause unless he determines that sufficient new information is available to warrant such proposal.

(iii) If the one-year period specified in subparagraph (A) is extended under clause (i) with respect to a proposed regulation, then before the close of such extended period the Secretary shall publish in the Federal Register either a final regulation to implement the determination or revision concerned, a finding that the revision should not be made, or a notice of withdrawal of the regulation under clause (ii), together with the finding on which the withdrawal is based.

(C) A final regulation designating critical habitat of an endangered species or a threatened species shall be published concurrently with the final regulation implementing the determination that such species is endangered or threatened, unless the Secretary deems that—

(i) it is essential to the conservation of such species that the regulation implementing such determination be promptly published; or

(ii) critical habitat of such species is not then determinable, in which case the Secretary, with respect to the proposed regulation to designate such habitat, may extend the one-year period specified in subparagraph (A) by not more than one additional year, but not later than the close of such additional year the Secretary must publish a final regulation, based on such data as may be available at that time, designating, to the maximum extent prudent, such habitat.

(7) Neither paragraph (4), (5), or (6) of this subsection nor section 553 of title 5, United States Code, shall apply to any regulation issued by the Secretary in regard to any emergency posing a significant risk to the well-being of any species of fish and wildlife or plants, but only if—

(A) at the time of publication of the regulation in the Federal Register the Secretary publishes therein detailed reasons why such regulation is necessary; and

(B) in the case such regulation applies to resident species of fish or wildlife, or plants, the Secretary gives actual notice of such regulation to the State agency in each State in which such species is believed to occur.

Such regulation shall, at the discretion of the Secretary, take effect immediately upon the publication of the regulation in the Federal Register. Any regulation promulgated under the authority of this paragraph shall cease to have force and effect at the close of the 240-day period following the date of publication unless, during such 240-day period, the rulemaking procedures which would apply to such regulation without regard to this paragraph are complied with. If at any time after issuing an emergency regulation the Secretary determines, on the basis of the best appropriate data available to him, that substantial evidence does not exist to warrant such regulation, he shall withdraw it.

(8) The publication in the Federal Register of any proposed or final regulation which is necessary or appropriate to carry out the purposes of this Act shall include a summary by the Secretary of the data on which such regulation is based and shall show the relationship of such data to such regulation; and if such regulation designates or revises critical habitat, such summary shall, to the maximum extent practicable, also include a brief description and evaluation of those activities (whether public or private) which, in the opinion of the Secretary, if undertaken may adversely modify such habitat, or may be affected by such designation.

(c) LISTS.—(1) The Secretary of the Interior shall publish in the Federal Register a list of all species determined by him or the Secretary of Commerce to be endangered species and a list of all species determined by him or the Secretary of Commerce to be threatened species. Each list shall refer to the species contained therein by scientific and common name or names, if any, specify with respect to such species over what portion of its range it is endangered or threatened, and specify any critical habitat within such range. The Secretary shall from time to time revise each list published under the authority of this subsection to reflect recent determinations, designations, and revisions made in accordance with subsections (a) and (b).

(2) The Secretary shall—

    (A) conduct, at least once every five years, a review of all species included in a list which is published pursuant to paragraph (1) and which is in effect at the time of such review; and

    (B) determine on the basis of such review whether any such species should—

        (i) be removed from such list;

        (ii) be changed in status from an endangered species to a threatened species; or

        (iii) be changed in status from a threatened species to an endangered species.

Each determination under subparagraph (B) shall be made in accordance with the provisions of subsection (a) and (b).

(d) PROTECTIVE REGULATIONS.—Whenever any species is listed as a threatened species pursuant to subsection (c) of this section, the Secretary shall issue such regulations as he deems necessary and advisable to provide for the conservation of such species. The Secretary may by regulation prohibit with respect to any threatened species any act prohibited under section 9(a)(1), in the case of fish or wildlife, or section 9(a)(2), in the case of plants, with respect to endangered species; except that with respect to the taking of resident species of fish or wildlife, such regulations shall apply in any State which has entered into a cooperative agreement pursuant to section 6(c) of this Act only to the extent that such regulations have also been adopted by such State.

(e) SIMILARITY OF APPEARANCE CASES.—The Secretary may, by regulation of commerce or taking, and to the extent he deems advisable, treat any species as an endangered species or threatened species even though it is not listed pursuant to section 4 of this Act if he finds that—

    (A) such species so closely resembles in appearance, at the point in question, a species which has been listed pursuant to such section that enforcement personnel would have substantial difficulty in attempting to differentiate between the listed and unlisted species;

    (B) the effect of this substantial difficulty is an additional threat to an endangered or threatened species; and

    (C) such treatment of an unlisted species will substantially facilitate the enforcement and further the policy of this Act.

    (f)(1) RECOVERY PLANS.—The Secretary shall develop and implement plans (hereinafter in this subsection referred to as 'recovery plans') for the conservation and survival of endangered species and threatened species listed pursuant to this section, unless he finds that such a plan will not promote the conservation of the species. The Secretary, in development and implementing recovery plans, shall, to the maximum extent practicable—

    (A) give priority to those endangered species or threatened species, without regard to taxonomic classification, that are most likely to benefit from such plans, particularly those species that are, or may be, in conflict with construction or other development projects or other forms of economic activity;

    (B) incorporate in each plan—

        (i) a description of such site-specific management actions as may be necessary to achieve the plan's goal for the conservation and survival of the species;

        (ii) objective, measurable criteria which, when met, would result in a determination, in accordance with the provisions of this section, that the species be removed from the list; and

        (iii) estimates of the time required and the cost to carry out those measures needed to achieve the plan's goal and to achieve intermediate steps toward that goal.

    (2) The Secretary, in developing and implementing recovery plans, may procure the services of appropriate public and private agencies and institutions, and other qualified persons. Recovery teams appointed pursuant to this subsection shall not be subject to the Federal Advisory Committee Act.

    (3) The Secretary shall report every two years to the Committee on Environment and Public Works of the Senate and the Committee on Merchant Marine and Fisheries of the House of Representatives on the status of efforts to develop and implement recovery plans for all species listed pursuant to this section and on the status of all species for which such plans have been developed.

    (4) The Secretary shall, prior to final approval of a new or revised recovery plan, provide public notice and an opportunity for public review and comment on such plan. The Secretary shall consider all information presented during the public comment period prior to approval of the plan.

    (5) Each Federal agency shall, prior to implementation of a new or revised recovery plan, consider all information presented during the public comment period under paragraph (4).

(g) MONITORING.—(1) The Secretary shall implement a system in cooperation with the States to monitor effectively for not less than five years the status of all species which have recovered to the point at which the measures provided pursuant to this Act are no longer necessary and which, in accordance with the provisions of this section, have been removed from either of the lists published under subsection (c).

(2) The Secretary shall make prompt use of the authority under paragraph 7 of subsection (b) of this section to prevent a significant risk to the well being of any such recovered species.

(h) AGENCY GUIDELINES.—The Secretary shall establish, and publish in the Federal Register, agency guidelines to insure that the purposes of this section are achieved efficiently and effectively. Such guidelines shall include, but are not limited to—

(1) procedures for recording the receipt and the disposition of petitions submitted under subsection (b)(3) of this section;

(2) criteria for making the findings required under such subsection with respect to petitions;

(3) a ranking system to assist in the identification of species that should receive priority review under subsection (a)(1) of the section; and

(4) a system for developing and implementing, on a priority basis, recovery plans under subsection (f) of this section. The Secretary shall provide to the public notice of, and opportunity to submit written comments on, any guideline (including any amendment thereto) proposed to be established under this subsection.

(i) If, in the case of any regulation proposed by the Secretary under the authority of this section, a State agency to which notice thereof was given in accordance with subsection (b)(5)(A)(ii) files comments disagreeing with all or part of the proposed regulation, and the Secretary issues a final regulation which is in conflict with such comments, or if the Secretary fails to adopt a regulation pursuant to an action petitioned by a State agency under subsection (b)(3), the Secretary shall submit to the State agency a written justification for his failure to adopt regulations consistent with the agency's comments or petition.

LAND ACQUISITION

SEC. 5. (a) PROGRAM.—The Secretary, and the Secretary of Agriculture with respect to the National Forest System, shall establish and implement a program to conserve fish, wildlife, and plants, including those which are listed as endangered species or threatened species pursuant to section 4 of this Act. To carry out such a program, the appropriate Secretary—

(1) shall utilize the land acquisition and other authority under the Fish and Wildlife Act of 1956, as amended, the Fish and Wildlife Coordination Act, as amended, and the Migratory Bird Conservation Act, as appropriate; and

(2) is authorized to acquire by purchase, donation, or otherwise, lands, waters, or interest therein, and such authority shall be in addition to any other land acquisition vested in him.

(b) ACQUISITIONS.—Funds made available pursuant to the Land and Water Conservation Fund Act of 1965, as amended, may be used for the purpose of acquiring lands, waters, or interests therein under subsection (a) of this section.

COOPERATION WITH THE STATES

SEC. 6. (a) GENERAL.—In carrying out the program authorized by this Act, the Secretary shall cooperate to the maximum extent practicable with the States. Such cooperation shall include consultation with the States concerned before acquiring any land or water, or interest therein, for the purpose of conserving any endangered species or threatened species.

(b) MANAGEMENT AGREEMENTS.—The Secretary may enter into agreements with any State for the administration and management of any area established for the conservation of endangered species or threatened species. Any revenues derived from the administration of such areas under these agreements shall be subject to the provisions of section 401 of the Act of June 15, 1935 (49 Stat. 383; 16 U.S.C. 715s).

(c)(1) COOPERATIVE AGREEMENTS.—In furtherance of the purposes of this Act, the Secretary is authorized to enter into a cooperative agreement in accordance with this section with any State which establishes and maintains an adequate and active program for the conservation of endangered species and threatened species. Within one hundred and twenty days after the Secretary receives a certified copy of such a proposed State program, he shall make a determination whether such program is in accordance with this Act. Unless he determines, pursuant to this paragraph, that the State program is not in accordance with this Act, he shall enter into a cooperative agreement with the State for the purpose of assisting in implementation of the State program. In order for a State program to be deemed an adequate and active program for the conservation of endangered species and threatened species, the Secretary must find, and annually thereafter reconfirm such finding, that under the State program—

(A) authority resides in the State agency of conserve resident species of fish or wildlife determined by the State agency or the Secretary to be endangered or threatened;

(B) the State agency has established acceptable conservation programs, consistent with the purposes and policies of this Act, for all resident species of fish or wildlife in the State which are deemed by the Secretary to be endangered or threatened, and has furnished a copy of such plan and program together with all pertinent details, information, and data requested to the Secretary;

(C) the State agency is authorized to conduct investigations to determine the status and requirements for survival of resident species of fish and wildlife;

(D) the State agency is authorized to establish programs, including the acquisition of land or aquatic habitat or interests therein, for the conservation of resident endangered or threatened species of fish or wildlife; and

(E) provision is made for public participation in designating resident species of fish or wildlife as endangered or threatened, or that under the State program—

(i) the requirements set forth in paragraphs (3), (4), and (5) of this subsection are complied with, and

(ii) plans are included under which immediate attention will be given to those resident species of fish and wildlife which are determined by the Secretary or the State agency to be endangered or threatened and which the Secretary and the State agency agree are most urgently in need of conservation programs; except that a cooperative agreement entered into with a State whose program is deemed adequate and active pursuant to clause (i) and this clause and this subparagraph shall not affect the applicability of prohibitions set forth in or authorized pursuant to section 4(d) or section 9(a)(1) with respect to the taking of any resident endangered or threatened species.

(2) In furtherance of the purposes of this Act, the Secretary is authorized to enter into a cooperative agreement in accordance with this section with any State which establishes and maintains an adequate and active program for the conservation of endangered species and threatened species of plants. Within one hundred and twenty days after the Secretary receives a certified copy of such a proposed State program, he shall make a determination whether such program is in accordance with this Act. Unless he determines, pursuant to this paragraph, that the State program is not in accordance with this Act, he shall enter into a cooperative agreement with the State for the purpose of assisting in implementation of the State program. In order for a State program to be deemed an adequate and active program for the conservation of endangered species of plants and threatened species of plants, the Secretary must find, and annually thereafter reconfirm such finding, that under the State program—

(A) authority resides in the State agency to conserve resident species of plants determined by the State agency or the Secretary to be endangered or threatened;

(B) the State agency has established acceptable conservation programs, consistent with the purposes and policies of this Act, for all resident species of plants in the State which are deemed by the Secretary to be endangered or threatened, and has furnished a copy of such plan and program together with all pertinent details, information, and data requested to the Secretary;

(C) the State agency is authorized to conduct investigations to determine the status and requirements for survival of resident species of plants; and

(D) provision is made for public participation in designating resident species of plants as endangered or threatened; or that under the State program—

(i) the requirements set forth in subparagraphs (C) and (D) of this paragraph are complied with, and

(ii) plans are included under which immediate attention will be given to those resident species of plants which are determined by the Secretary or the State agency to be endangered or threatened and which the Secretary and the State agency agree are most urgently in need of conservation programs; except that a cooperative agreement entered into with a State whose program is deemed adequate and active pursuant to clause (i) and this clause shall not affect the applicability of prohibitions set forth in or authorized pursuant to section 4(d) or section 9(a)(1) with respect to the taking of any resident endangered or threatened species.

(d) ALLOCATION OF FUNDS.—(1) The Secretary is authorized to provide financial assistance to any State, through its respective State agency, which has entered into a cooperative agreement pursuant to subsection (c) of this section to assist in development of programs for the conservation of endangered and threatened species or to assist in monitoring the status of candidate species pursuant to subparagraph (C) of section 4(b)(3) and recovered species pursuant to section 4(g). The Secretary shall allocate each annual appropriation made in accordance with the provisions of subsection (i) of this section to such States based on consideration of—

(A) the international commitments of the United States to protect endangered species or threatened species;

(B) the readiness of a State to proceed with a conservation program consistent with the objectives and purposes of this Act;

(C) the number of endangered species and threatened species within a State;

(D) the potential for restoring endangered species and threatened species within a State;

(E) the relative urgency to initiate a program to restore and protect an endangered species or threatened species in terms of survival of the species;

(F) the importance of monitoring the status of candidate species within a State to prevent a significant risk to the well being of any such species; and

(G) the importance of monitoring the status of recovered species within a State to assure that such species do not return to the point at which the measures provided pursuant to this Act are again necessary.

So much of the annual appropriation made in accordance with provisions of subsection (i) of this section allocated for obligation to any State for any fiscal year as remains unobligated at the close thereof is authorized to be made available to that State until the close of the succeeding fiscal year. Any amount allocated to any State which is unobligated at the end of the period during which it is available for expenditure is authorized to be made available for expenditure by the Secretary in conducting programs under this section.

(2) Such cooperative agreements shall provide for (A) the actions to be taken by the Secretary and the States; (B) the benefits that are expected to be derived in connection with the conservation of endangered or threatened species; (C) the estimated cost of these actions; and (D) the share of such costs to be borne by the Federal Government and by the States; except that—

(i) the Federal share of such program costs shall not exceed 75 percent of the estimated program cost stated in the agreement; and

(ii) the Federal share may be increased to 90 percent whenever two or more States having a common interest in one or more endangered or threatened species, the conservation of which may be enhanced by cooperation of such States, enter jointly into agreement with the Secretary.

The Secretary may, in his discretion, and under such rules and regulations as he may prescribe, advance funds to the State for financing the United States pro rata share agreed upon in the cooperative agreement. For the purposes of this section, the non-Federal share may, in the discretion of the Secretary, be in the form of money or real property, the value of which will be determined by the Secretary whose decision shall be final.

(e) REVIEW OF STATE PROGRAMS.—Any action taken by the Secretary under this section shall be subject to his periodic review at no greater than annual intervals.

(f) CONFLICTS BETWEEN FEDERAL AND STATE LAWS.—Any State law or regulation which applies with respect to the importation or exportation of, or interstate or foreign commerce in, endangered species or threatened species is void to the extent that it may effectively (1) permit what is prohibited by this Act of by any regulation which implements this Act, or (2) prohibit what is authorized pursuant to an exemption or permit provided for in this Act or in any regulation which implements this Act. This Act shall not otherwise be construed to void any State law or regulation which is intended to conserve migratory, resident, or introduced fish or wildlife, or to permit or prohibit sale of such fish or wildlife. Any State law or regulation respecting the taking of an endangered species or threatened species may be more restrictive than the exemptions or permits provided for in this Act or in any regulation which implements this Act but not less restrictive than the prohibitions so defined.

(g) TRANSITION.—(1) For purposes of this subsection, the term "establishment period" means, with respect to any State, the period beginning on the date of enactment of this Act and ending on whichever of the following dates first occurs: (A) the date of the close of the 120-day period following the adjournment of the first regular session of the legislature of such State which commences after such date of enactment, or (B) the date of the close of the 15-month period following such date of enactment.

(2) The prohibitions set forth in or authorized pursuant to sections 4(d) and 9(a)(1)(B) of this Act shall not apply with respect to the taking of any resident endangered species or threatened species (other than species listed in Appendix I to the Convention or otherwise specifically covered by any other treaty or Federal law) within any State—

(A) which is then a party to a cooperative agreement with the Secretary pursuant to section 6(c) of this Act (except to the extent that the taking of any such species is contrary to the law of such State); or

(B) except for any time within the establishment period when—

(i) the Secretary applies such prohibition to such species at the request of the State, or

(ii) the Secretary applies such prohibition after he finds, and publishes his finding, that an emergency exists posing a significant risk to the well-being of such species and that the prohibition must be applied to protect such species. The Secretary's finding and publication may be made without regard to the public hearing or comment provisions of section 553 of title 5, United States Code, or any other provision of this Act; but such prohibition shall expire 90 days after the date of its imposition unless the Secretary further extends such prohibition by publishing notice and a statement of justification of such extension.

(h) REGULATIONS.—The Secretary is authorized to promulgate such regulations as may be appropriate to carry out the provisions of this section relating to financial assistance to States.

(i) APPROPRIATIONS.—(1) To carry out the provisions of this section for fiscal years after September 30, 1988, there shall be deposited into a special fund known as the cooperative endangered species conservation fund, to be administered by the Secretary, an amount equal to five percent of the combined amounts covered each fiscal year into the Federal aid to wildlife restoration fund under section 3 of the Act of September 2, 1937, and paid, transferred, or otherwise credited each fiscal year to the Sport Fishing Restoration Account established under 1016 of the Act of July 18, 1984.

(2) Amounts deposited into the special fund are authorized to be appropriated annually and allocated in accordance with subsection (d) of this section.

INTERAGENCY COOPERATION

SEC. 7. (a) FEDERAL AGENCY ACTIONS AND CONSULTATIONS.—(1) The Secretary shall review other programs administered by him and utilize such programs in furtherance of the purposes of this Act. All other Federal agencies shall, in consultation with and with the assistance of the Secretary, utilize their authorities in furtherance of the purposes of this Act by carrying out programs for the conservation of endangered species and threatened species listed pursuant to section 4 of this Act.

(2) Each Federal agency shall, in consultation with and with the assistance of the Secretary, insure that any action authorized, funded, or carried out by such agency (hereinafter in this section referred to as an "agency action") is not likely to jeopardize the continued existence of any endangered species or threatened species or result in the destruction or adverse modification of habitat of such species which is determined by the Secretary, after consultation as appropriate with affected States, to be critical, unless such agency has been granted an exemption for such action by the Committee pursuant to subsection (h) of this section. In fulfilling the requirements of this paragraph each agency shall use the best scientific and commercial data available.

(3) Subject to such guidelines as the Secretary may establish, a Federal agency shall consult with the Secretary on any prospective

agency action at the request of, and in cooperation with, the prospective permit or license applicant if the applicant has reason to believe that an endangered species or a threatened species may be present in the area affected by his project and that implementation of such action will likely affect such species.

(4) Each Federal agency shall confer with the Secretary on any agency action which is likely to jeopardize the continued existence of any species proposed to be listed under section 4 or result in the destruction or adverse modification of critical habitat proposed to be designated for such species. This paragraph does not require a limitation on the commitment of resources as described in subsection (d).

(b) OPINION OF SECRETARY.—(1)(A) Consultation under subsection (a)(2) with respect to any agency action shall be concluded within the 90-day period beginning on the date on which initiated or, subject to subparagraph (B), within such other period of time as is mutually agreeable to the Secretary and the Federal agency;

(B) in the case of an agency action involving a permit or license applicant, the Secretary and the Federal agency may not mutually agree to conclude consultation within a period exceeding 90 days unless the Secretary, before the close of the 90th day referred to in subparagraph (A)—

(i) if the consultation period proposed to be agreed to will end before the 150th day after the date on which consultation was initiated, submits to the applicant a written statement setting forth—

(I) the reasons why a longer period is required;

(II) the information that is required to complete the consultation; and

(III) the estimated date on which consultation will be completed; or

(ii) if the consultation period proposed to be agreed to will end 150 or more days after the date on which consultation was initiated, obtains the consent of the applicant to such period.
The Secretary and the Federal agency may mutually agree to extend a consultation period established under the preceding sentence if the Secretary, before the close of such period, obtains the consent of the applicant to the extension.

(2) Consultation under subsection (a)(3) shall be concluded within such period as is agreeable to the Secretary, the Federal agency, and the applicant concerned.

(3)(A) Promptly after conclusion of consultation under paragraph (2) or (3) of subsection (a), the Secretary shall provide to the Federal agency and the applicant, if any, a written statement setting forth the Secretary's opinion, and a summary of the information on which the opinion is based, detailing how the agency action affects the species or its critical habitat. If jeopardy or adverse modification is found, the Secretary shall suggest those reasonable and prudent alternatives which he believes would not violate subsection (a)(2) and can be taken by the Federal agency or applicant in implementing the agency action.

(B) Consultation under subsection (a)(3), and an opinion based by the Secretary incident to such consultation, regarding an agency action shall be treated respectively as a consultation under subsection (a)(2), and as an opinion issued after consultation under such subsection, regarding that action if the Secretary reviews the action before it is commenced by the Federal agency and finds, and notifies such agency, that no significant changes have been made with respect to the action and that no significant change has occurred regarding the information used during the initial consultation.

(4) If after consultation under subsection (a)(2) of this section, the Secretary concludes that—

(A) the agency action will not violate such subsection, or offers reasonable and prudent alternatives which the Secretary believes would not violate such subsection;

(B) the taking of an endangered species or a threatened species incidental to the agency action will not violate such subsection; and

(C) if an endangered species or threatened species of a marine mammal is involved, the taking is authorized pursuant to section 1371(a)(5) of this title;

the Secretary shall provide the Federal agency and the applicant concerned, if any, with a written statement that—

(i) specifies the impact of such incidental taking on the species,

(ii) specifies those reasonable and prudent measures that the Secretary considers necessary or appropriate to minimize such impact,

(iii) in the case of marine mammals, specifies those measures that are necessary to comply with section 1371(a)(5) of this title with regard to such taking, and

(iv) sets forth the terms and conditions (including, but not limited to, reporting requirements) that must be complied with by the Federal agency or applicant (if any), or both, to implement the measures specified under clauses (ii) and (iii).

(c) BIOLOGICAL ASSESSMENT.—(1) To facilitate compliance with the requirements of subsection (a)(2) each Federal agency shall, with respect to any agency action of such agency for which no contract for construction has been entered into and for which no construction has begun on the date of enactment of the Endangered Species Act Amendments of 1978, request of the Secretary information whether any species which is listed or proposed to be listed may be present in the area of such proposed action. If the Secretary advises, based on the best scientific and commercial data available, that such species may be present, such agency shall conduct a biological assessment for the purpose of identifying any endangered species or threatened species which is likely to be affected by such action. Such assessment shall be completed within 180 days after the date on which initiated (or within such other period as is mutually agreed to by the Secretary and such agency, except that if a permit or license applicant is involved, the 180-day period may not be extended unless such agency provides the applicant, before the close of such period, with a written statement setting forth the estimated length of the proposed extension and the reasons therefor) and, before any contract for construction is entered into and before construction is begun with respect to such action. Such assessment may be undertaken as part of a Federal agency's compliance with the requirements of section 102 of the National Environmental Policy Act of 1969 (42 U.S.C. 4332).

(2) Any person who may wish to apply for an exemption under subsection (g) of this section for that action may conduct a biological assessment to identify any endangered species or threatened species which is likely to be affected by such action. Any such biological assessment must, however, be conducted in cooperation with the Secretary and under the supervision of the appropriate Federal agency.

(d) LIMITATION ON COMMITMENT OF RESOURCES.—After initiation of consultation required under subsection (a)(2), the Federal agency and the permit or license applicant shall not make any irreversible or irretrievable commitment of resources with respect to the agency action which has the effect of foreclosing the formulation or implementation of any reasonable and prudent alternative measures which would not violate subsection (a)(2).

(e)(1) ESTABLISHMENT OF COMMITTEE.—There is established a committee to be known as the Endangered Species Committee (hereinafter in this section referred to as the "Committee").

(2) The Committee shall review any application submitted to it pursuant to this section and determine in accordance with subsection (h) of this section whether or not to grant an exemption from the requirements of subsection (a)(2) of this action for the action set forth in such application.

(3) The Committee shall be composed of seven members as follows:

(A) The Secretary of Agriculture.

(B) The Secretary of the Army.

(C) The Chairman of the Council of Economic Advisors.

(D) The Administrator of the Environmental Protection Agency. Agency.

(E) The Secretary of the Interior.

(F) The Administrator of the National Oceanic and Atmospheric Administration.

(G) The President, after consideration of any recommendations received pursuant to subsection (g)(2)(B) shall appoint one individual from each affected State, as determined by the Secretary, to be a member of the Committee for the consideration of the application for exemption for an agency action with respect to which such recommendations are made, not later than 30 days after an application is submitted pursuant to this section.

(4)(A) Members of the Committee shall receive no additional pay on account of their service on the Committee.

(B) While away from their homes or regular places of business in the performance of services for the Committee, members of the Committee shall be allowed travel expenses, including per diem in lieu of subsistence, in the same manner as persons employed intermittently in the Government service are allowed expenses under section 5703 of title 5 of the United States Code

(5)(A) Five members of the Committee or their representatives shall constitute a quorum for the transaction of any function of the Committee, except that, in no case shall any representative be considered in determining the existence of a quorum for the transaction of any function of the Committee if that function involves a vote by the Committee on any matter before the Committee.

(B) The Secretary of the Interior shall be the Chairman of the Committee.

(C) The Committee shall meet at the call of the Chairman or five of its members.

(D) All meetings and records of the Committee shall be open to the public.

(6) Upon request of the Committee, the head of any Federal agency is authorized to detail, on a nonreimbursable basis, any of the personnel of such agency to the Committee to assist it in carrying out its duties under this section.

(7)(A) The Committee may for the purpose of carrying out its duties under this section hold such hearings, sit and act at such times and places, take such testimony, and receive such evidence, as the Committee deems advisable.

(B) When so authorized by the Committee, any member or agent of the Committee may take any action which the Committee is authorized to take by this paragraph.

(C) Subject to the Privacy Act, the Committee may secure directly from any Federal agency information necessary to enable it to carry out its duties under this section. Upon request of the Chairman of the Committee, the head of such Federal agency shall furnish such information to the Committee.

(D) The Committee may use the United States mails in the same manner and upon the same conditions as a Federal agency.

(E) The Administrator of General Services shall provide to the Committee on a reimbursable basis such administrative support services as the Committee may request.

(8) In carrying out its duties under this section, the Committee may promulgate and amend such rules, regulations, and procedures, and issue and amend such orders as it deems necessary.

(9) For the purpose of obtaining information necessary for the consideration of an application for an exemption under this section the Committee may issue subpoenas for the attendance and testimony of witnesses and the production of relevant papers, books, and documents.

(10) In no case shall any representative, including a representative of a member designated pursuant to paragraph (3)(G) of this subsection, be eligible to cast a vote on behalf of any member.

(f) REGULATIONS.—Not later than 90 days after the date of enactment of the Endangered Species Act Amendments of 1978, the Secretary shall promulgate regulations which set forth the form and manner in which applications for exemption shall be submitted to the Secretary and the information to be contained in such applications. Such regulations shall require that information submitted in an application by the head of any Federal agency with respect to any agency action include but not be limited to—

(1) a description of the consultation process carried out pursuant to subsection (a)(2) of this section between the head of the Federal agency and the Secretary; and

(2) a statement describing why such action cannot be altered or modified to conform with the requirements of subsection (a)(2) of this section.

(g) APPLICATION FOR EXEMPTION AND REPORT TO THE COMMITTEE.—(1) A Federal agency, the Governor of the State in which an agency action will occur, if any, or a permit or license applicant may apply to the Secretary for an exemption for an agency action of such agency if, after consultation under subsection (a)(2), the Secretary's opinion under subsection (b) indicates that the agency action would violate subsection (a)(2). An application for an exemption shall be considered initially by the Secretary in the manner provided for in this subsection, and shall be considered by the Committee for a final determination under subsection (h) after a report is made pursuant to paragraph (5). The applicant for an exemption shall be referred to as the "exemption applicant" in this section.

(2)(A) An exemption applicant shall submit a written application to the Secretary, in a form prescribed under subsection (f), not later than 90 days after the completion of the consultation process; except that, in the case of any agency action involving a permit or license applicant, such application shall be submitted not later than 90 days after the date on which the Federal agency concerned takes final agency action with respect to the issuance of the permit or license. For purposes of the preceding sentence, the term "final agency action" means (i) a disposition by an agency with respect to the issuance of a permit or license that is subject to administrative review, whether or not such disposition is subject to judicial review; or (ii) if administrative review is sought with respect to such disposition, the decision resulting after such review. Such application shall set forth the reasons why the exemption applicant considers that the agency action meets the requirements for an exemption under this subsection.

(B) Upon receipt of an application for exemption for an agency action under paragraph (1), the Secretary shall promptly (i) notify the Governor of each affected State, if any, as determined by the Secretary, and request the Governors so notified to recommend individuals to be appointed to the Endangered Species Committee for consideration of such application; and (ii) publish notice of receipt of the application in the Federal Register, including a summary of the information contained in the application and a description of the agency action with respect to which the application for exemption has been filed.

(3) The Secretary shall within 20 days after the receipt of an application for exemption, or within such other period of time as is mutually agreeable to the exemption applicant and the Secretary—

(A) determine that the Federal agency concerned and the exemption applicant have—

(i) carried out the consultation responsibilities described in subsection (a) in good faith and made a reasonable and responsible effort to develop and fairly consider modifications or reasonable and prudent alternatives to the proposed agency action which would not violate subsection (a)(2);

(ii) conducted any biological assessment required by subsection (c); and

(iii) to the extent determinable within the time provided herein, refrained from making any irreversible or irretrievable commitment of resources prohibited by subsection (d); or

(B) deny the application for exemption because the Federal agency concerned or the exemption applicant have not met the requirements set forth in subparagraph (A) (i), (ii), and (iii).

The denial of an application under subparagraph (B) shall be considered final agency action for purposes of chapter 7 of title 5, United States Code.

(4) If the Secretary determines that the Federal agency concerned and the exemption applicant have met the requirements set forth in paragraph (3)(A) (i), (ii) and (iii) he shall, in consultation with the Members of the Committee, hold a hearing on the application for exemption in accordance with sections 554, 555, and 556 (other than subsection (b) (1) and (2) thereof) of title 5, United States Code, and prepare the report to be submitted pursuant to paragraph (5).

(5) Within 140 days after making the determinations under paragraph (3) or within such other period of time as is mutually agreeable to the exemption applicant and the Secretary, the Secretary shall submit to the Committee a report discussing—

(A) the availability of reasonable and prudent alternatives to the agency action, and the nature and extent of the benefits of the agency action and of alternative courses of action consistent with conserving the species of the critical habitat;

(B) a summary of the evidence concerning whether or not the agency action is in the public interest and is of national or regional significance;

(C) appropriate reasonable mitigation and enhancement measures which should be considered by the Committee; and

(D) whether the Federal agency concerned and the exemption applicant refrained from making any irreversible or irretrievable commitment of resources prohibited by subsection (d).

(6) To the extent practicable within the time required for action under subsection (g) of this section, and except to the extent inconsistent with the requirements of this section, the consideration of any application for an exemption under this section and the conduct of any hearing under this subsection shall be in accordance with sections 554, 555, and 556 (other than subsection (b)(3) of section 556) of title 5, United States Code.

(7) Upon request of the Secretary, the head of any Federal agency is authorized to detail, on a nonreimbursable basis, any of the personnel of such agency to the Secretary to assist him in carrying out his duties under this section.

(8) All meetings and records resulting from activities pursuant to this subsection shall be open to the public.

(h) EXEMPTION.—(1) The Committee shall make a final determination whether or not to grant an exemption within 30 days after receiving the report of the Secretary pursuant to subsection (g)(5). The Committee shall grant an exemption from the requirements of subsection (a)(2) for an agency action if, by a vote of not less than five of its members voting in person—

(A) it determines on the record, based on the report of the Secretary, the record of the hearing held under subsection (g)(4), and on such other testimony or evidence as it may receive, that—

(i) there are no reasonable and prudent alternatives to the agency action;

(ii) the benefits of such action clearly outweigh the benefits of alternative courses of action consistent with conserving the species or its critical habitat, and such action is in the public interest;

(iii) the action is of regional or national significance; and

(iv) neither the Federal agency concerned nor the exemption applicant made any irreversible or irretrievable commitment of resources prohibited by subsection (d); and

(B) it establishes such reasonable mitigation and enhancement measures, including, but not limited to, live propagation, transplantation, and habitat acquisition and improvement, as are necessary and appropriate to minimize the adverse effects of the agency action upon the endangered species, threatened species, or critical habitat concerned.

Any final determination by Committee under this subsection shall be considered final agency action for purposes of chapter 7 of title 5 of the United States Code.

(2)(A) Except as provided in subparagraph (B), an exemption for an agency action granted under paragraph (1) shall constitute a permanent exemption with respect to all endangered or threatened species for the purposes of completing such agency action—

(i) regardless whether the species was identified in the biological assessment; and

(ii) only if a biological assessment has been conducted under subsection (c) with respect to such agency action.

(B) An exemption shall be permanent under subparagraph (A) unless—

(i) the Secretary finds, based on the best scientific and commercial data available, that such exemption would result in the extinction of a species that was not the subject of consultation under subsection (a)(2) or was not identified in any biological assessment conducted under subsection (c), and

(ii) the Committee determines within 60 days after the date of the Secretary's finding that the exemption should not be permanent.

If the Secretary makes a finding described in clause (i), the Committee shall meet with respect to the matter within 30 days after the date of the finding.

(i) REVIEW BY SECRETARY OF STATE.—Notwithstanding any other provision of this Act, the Committee shall be prohibited from considering for exemption any application made to it, if the Secretary of State, after a review of the proposed agency action and its potential implications, and after hearing, certifies, in writing, to the Committee within 60 days of any application made under this section that the granting of any such exemption and the carrying out of such action would be in violation of an international treaty obligation or other international obligation of the United States. The Secretary of State shall, at the time of such certification, publish a copy thereof in the Federal Register.

(j) Notwithstanding any other provision of this Act, the Committee shall grant an exemption for any agency action if the Secretary of Defense finds that such exemption is necessary for reasons of national security.

(k) SPECIAL PROVISIONS.—An exemption decision by the Committee under this section shall not be a major Federal action for purposes of the National Environmental Policy Act of 1969 (42 U.S.C. 4321 et seq.): *Provided,* That an environmental impact statement which discusses the impacts upon endangered species or threatened species or their critical habitats shall have been previously prepared with respect to any agency action exempted by such order.

(l) COMMITTEE ORDERS.—(1) If the Committee determines under subsection (h) that an exemption should be granted with respect to any agency action, the Committee shall issue an order granting the exemption and specifying the mitigation and enhancement measures established pursuant to subsection (h) which shall be carried out and paid for by the exemption applicant in implementing the agency action. All necessary mitigation and enhancement measures shall be authorized prior to the implementing of the agency action and funded concurrently with all other project features.

(2) The applicant receiving such exemption shall include the costs of such mitigation and enhancement measures within the overall costs of continuing the proposed action. Notwithstanding the preceding sentence the costs of such measures shall not be treated as project costs for the purpose of computing benefit-cost or other ratios for the proposed action. Any applicant may request the Secretary to carry out such mitigation and enhancement measures. The costs incurred by the Secretary in carrying out any such measures shall be paid by the applicant receiving the exemption. No later than one year after the granting of an exemption, the exemption applicant shall submit to the Council on Environmental Quality a report describing its compliance with the mitigation and enhancement measures prescribed by this section. Such report shall be submitted annually until all such mitigation and enhancement measures have been completed. Notice of the public availability of such reports shall be published in the Federal Register by the Council on Environmental Quality.

(m) NOTICE.—The 60-day notice requirement of section 11(g) of this Act shall not apply with respect to review of any final determination of the Committee under subsection (h) of this section granting an exemption from the requirements of subsection (a)(2) of this section.

(n) JUDICIAL REVIEW.—Any person, as defined by section 3(13) of this Act, may obtain judicial review, under chapter 7 of title 5 of the United States Code, of any decision of the Endangered Species Committee under subsection (h) in the United States Court of Appeals for (1) any circuit wherein the agency action concerned will be, or is being, carried out, or (2) in any case in which the agency action will be, or is being, carried out outside of any circuit, the District of Columbia, by filing in such court within 90 days after the date of issuance of the decision, a written petition for review. A copy of such petition shall be transmitted by the clerk of the court to the Committee and the Committee shall file in the court the record in the proceeding, as provided in section 2112, of title 28,

United States Code. Attorneys designated by the Endangered Species Committee may appear for, and represent the Committee in any action for review under this subsection.

(o) EXEMPTION AS PROVIDING EXCEPTION ON TAKING OF ENDANGERED SPECIES.—Notwithstanding sections 1533(d) and 1538(a)(1)(B) and (C) of this title, sections 1371 and 1372 of this title, or any regulation promulgated to implement any such section—

(1) any action for which an exemption is granted under subsection (h) of this section shall not be considered to be a taking of any endangered species or threatened species with respect to any activity which is necessary to carry out such action; and

(2) any taking that is in compliance with the terms and conditions specified in a written statement provided under subsection (b)(4)(iv) of this section shall not be considered to be a prohibited taking of the species concerned.

(p) EXEMPTIONS IN PRESIDENTIALLY DECLARED DISASTER AREAS.— In any area which has been declared by the President to be a major disaster area under the Disaster Relief Act of 1974, the President is authorized to make the determinations required by subsections (g) and (h) of this section for any project for the repair or replacement of a public facility substantially as it existed prior to the disaster under section 401 or 402 of the Disaster Relief Act of 1974, and which the President determines (1) is necessary to prevent the recurrence of such a natural disaster and to reduce the potential loss of human life, and (2) to involve an emergency situation which does not allow the ordinary procedures of this section to be followed. Notwithstanding any other provision of this section, the Committee shall accept the determinations of the President under this subsection.

INTERNATIONAL COOPERATION

SEC. 8. (a) FINANCIAL ASSISTANCE.—As a demonstration of the commitment of the United States to the worldwide protection of endangered species and threatened species, the President may, subject to the provisions of section 1415 of the Supplemental Appropriation Act, 1953 (31 U.S.C. 724), use foreign currencies accruing to the United States Government under the Agricultural Trade Development and Assistance Act of 1954 or any other law to provide to any foreign country (with its consent) assistance in the development and management of programs in that country which the Secretary determines to be necessary or useful for the conservation of any endangered species or threatened species listed by the Secretary pursuant to section 4 of this Act. The President shall provide assistance (which includes, but is not limited to, the acquisition, by lease or otherwise, of lands, waters, or interests therein) to foreign countries under this section under such terms and conditions as he deems appropriate. Whenever foreign currencies are available for the provision of assistance under this section, such currencies shall be used in preference to funds appropriated under the authority of section 15 of this Act.

(b) ENCOURAGEMENT OF FOREIGN PROGRAMS.—In order to carry out further the provisions of this Act, the Secretary, through the Secretary of State shall encourage—

(1) foreign countries to provide for the conservation of fish or wildlife and plants including endangered species and threatened species listed pursuant to section 4 of this Act;

(2) the entering into of bilateral or multilateral agreements with foreign countries to provide for such conservation; and

(3) foreign persons who directly or indirectly take fish or wildlife or plants in foreign countries or on the high seas for importation into the United States for commercial or other purposes to develop and carry out with such assistance as he may provide, conservation practices designed to enhance such fish or wildlife or plants and their habitat.

(c) PERSONNEL.—After consultation with the Secretary of State, the Secretary may—

(1) assign or otherwise make available any officer or employee of his department for the purpose of cooperating with foreign countries and international organizations in developing personnel resources and programs which promote the conservation of fish or wildlife or plants, and

(2) conduct or provide financial assistance for the educational training of foreign personnel, in this country or abroad, in fish, wildlife, or plant management, research and law enforcement and to render professional assistance abroad in such matters.

(d) INVESTIGATIONS.—After consultation with the Secretary of State and the Secretary of the Treasury, as appropriate, the Secretary may conduct or cause to be conducted such law enforcement investigations and research abroad as he deems necessary to carry out the purposes of this Act.

CONVENTION IMPLEMENTATION

SEC. 8A. (a) MANAGEMENT AUTHORITY AND SCIENTIFIC AUTHORITY.—The Secretary of the Interior (hereinafter in this section referred to as the "Secretary") is designated as the Management Authority and the Scientific Authority for purposes of the Convention and the respective functions of each such Authority shall be carried out through the United States Fish and Wildlife Service.

(b) MANAGEMENT AUTHORITY FUNCTIONS.—The Secretary shall do all things necessary and appropriate to carry out the functions of the Management Authority under the Convention.

(c) SCIENTIFIC AUTHORITY FUNCTIONS.—(1) The Secretary shall do all things necessary and appropriate to carry out the functions of the Scientific Authority under the Convention.

(2) The Secretary shall base the determinations and advice given by him under Article IV of the Convention with respect to wildlife upon the best available biological information derived from professionally accepted wildlife management practices; but is not required to make, or require any State to make, estimates of population size in making such determinations or giving such advice.

(d) RESERVATIONS BY THE UNITED STATES UNDER CONVENTION.—If the United States votes against including any species in Appendix I or II of the Convention and does not enter a reservation pursuant to paragraph (3) of Article XV of the Convention with respect to that species, the Secretary of State, before the 90th day after the last day on which such a reservation could be entered, shall submit to the Committee on Merchant Marine and Fisheries of the House of Representatives, and to the Committee on the Environment and Public Works of the Senate, a written report setting forth the reasons why such a reservation was not entered.

(e) WILDLIFE PRESERVATION IN WESTERN HEMISPHERE.—(1) The Secretary of the Interior (hereinafter in this subsection referred to as the "Secretary"), in cooperation with the Secretary of State, shall act on behalf of, and represent, the United States in all regards as required by the Convention on Nature Protection and Wildlife Preservation in the Western Hemisphere (56 Stat. 1354, T.S. 982, hereinafter in this subsection referred to as the "Western Convention"). In the discharge of these responsibilities, the Secretary and the Secretary of State shall consult with the Secretary of Agriculture, the Secretary of Commerce, and the heads of other agencies with respect to matters relating to or affecting their areas of responsibility.

(2) The Secretary and the Secretary of State shall, in cooperation with the contracting parties to the Western Convention and, to the extent feasible and appropriate, with the participation of State agencies, take such steps as are necessary to implement the Western Convention. Such steps shall include, but not be limited to—

(A) cooperation with contracting parties and international organizations for the purpose of developing personnel resources and programs that will facilitate implementation of the Western Convention;

(B) identification of those species of birds that migrate between the United States and other contracting parties, and the habitats upon which those species depend, and the implementation of cooperative measures to ensure that such species will not become endangered or threatened; and

(C) identification of measures that are necessary and appropriate to implement those provisions of the Western Convention which address the protection of wild plants.

(3) No later than September 30, 1985, the Secretary and the Secretary of State shall submit a report to Congress describing those steps taken in accordance with the requirements of this subsection and identifying the principal remaining actions yet necessary for comprehensive and effective implementation of the Western Convention.

(4) The provisions of this subsection shall not be construed as affecting the authority, jurisdiction, or responsibility of the several States to manage, control, or regulate resident fish or wildlife under State law or regulations.

PROHIBITED ACTS

SEC. 9. (a) GENERAL.—(1) Except as provided in sections 6(g)(2) and 10 of this Act, with respect to any endangered species of fish or wildlife listed pursuant to section 4 of this Act it is unlawful for any person subject to the jurisdiction of the United States to—

(A) import any such species into, or export any such species from the United States;

(B) take any such species within the United States or the territorial sea of the United States;

(C) take any such species upon the high seas;

(D) possess, sell, deliver, carry, transport, or ship, by any means whatsoever, any such species taken in violation of subparagraphs (B) and (C);

(E) deliver, receive, carry, transport, or ship in interstate or foreign commerce, by any means whatsoever and in the course of a commercial activity, any such species;

(F) sell or offer for sale in interstate or foreign commerce any such species; or

(G) violate any regulation pertaining to such species or to any threatened species of fish or wildlife listed pursuant to section 4 of this Act and promulgated by the Secretary pursuant to authority provided by this Act.

(2) Except as provided in sections 6(g)(2) and 10 of this Act, with respect to any endangered species of plants listed pursuant to section 4 of this Act, it is unlawful for any person subject to the jurisdiction of the United States to—

(A) import any such species into, or export any such species from, the United States;

(B) remove and reduce to possession any such species from areas under Federal jurisdiction; maliciously damage or destroy any such species on any such area; or remove, cut, dig up, or damage or destroy any such species on any other area in knowing violation of any law or regulation of any state or in the course of any violation of a state criminal trespass law;".

(C) deliver, receive, carry, transport, or ship in interstate or foreign commerce, by any means whatsoever and in the course of a commercial activity, any such species;

(D) sell or offer for sale in interstate or foreign commerce any such species; or

(E) violate any regulation pertaining to such species or to any threatened species of plants listed pursuant to section 4 of this Act and promulgated by the Secretary pursuant to authority provided by this Act.

(b)(1) SPECIES HELD IN CAPTIVITY OR CONTROLLED ENVIRONMENT.— The provisions of subsections (a)(1)(A) and (a)(1)(G) of this section shall not apply to any fish or wildlife which was held in captivity or in a controlled environment on (A) December 28, 1973, or (B) the date of the publication in the Federal Register of a final regulation adding such fish or wildlife species to any list published pursuant to subsection (c) of section 4 of this Act: *Provided,* That such holding and any subsequent holding or use of the fish or wildlife was not in the course of a commercial activity. With respect to any act prohibited by subsections (a)(1)(A) and (a)(1)(G) of this section which occurs after a period of 180 days from (i) December 28, 1973, or (ii) the date of publication in the Federal Register of a final regulation adding such fish or wildlife species to any list published pursuant to subsection (c) of section 4 of this Act, there shall be a rebuttable presumption that the fish or wildlife involved in such act is not entitled to the exemption contained in this subsection.

(2)(A) The provisions of subsections (a)(1) shall not apply to—

(i) any raptor legally held in captivity or in a controlled environment on the effective date of the Endangered Species Act Amendments of 1978; or

(ii) any progeny of any raptor described in clause (i); until such time as any such raptor or progeny is intentionally returned to a wild state.

(B) Any person holding any raptor or progeny described in subparagraph (A) must be able to demonstrate that the raptor or progeny does, in fact, qualify under the provisions of this paragraph, and shall maintain and submit to the Secretary, on request, such inventories, documentation, and records as the Secretary may by regulation require as being reasonably appropriate to carry out the purposes of this paragraph. Such requirements shall not unnecessarily duplicate the requirements of other rules and regulations promulgated by the Secretary.

(c) VIOLATION OF CONVENTION.—(1) It is unlawful for any person subject to the jurisdiction of the United States to engage in any trade in any specimens contrary to the provisions of the Convention, or to possess any specimens traded contrary to the provisions of the Convention, including the definitions of terms in article I thereof.

(2) Any importation into the United States of fish or wildlife shall, if—

(A) such fish or wildlife is not an endangered species listed pursuant to section 4 of this Act but is listed in Appendix II of the Convention;

(B) the taking and exportation of such fish or wildlife is not contrary to the provisions of the Convention and all other applicable requirements of the Convention have been satisfied;

(C) the applicable requirements of subsections (d), (e), and (f) of this section have been satisfied; and

(D) such importation is not made in the course of a commercial activity;

be presumed to be an importation not in violation of any provision of this Act or any regulation issued pursuant to this Act.

(d) IMPORTS AND EXPORTS.—

(1) IN GENERAL.—It is unlawful for any person, without first having obtained permission from the Secretary, to engage in business—

(A) as an importer or exporter of fish or wildlife (other than shellfish and fishery products which (i) are not listed pursuant to section 4 of this Act as endangered species or threatened species, and (ii) are imported for purposes of human or animal consumption or taken in waters under the jurisdiction of the United States or on the high seas for recreational purposes) or plants; or

(B) as an importer or exporter of any amount of raw or worked African elephant ivory.

(2) REQUIREMENTS.—Any person required to obtain permission under paragraph (1) of this subsection shall—

(A) keep such records as will fully and correctly disclose each importation or exportation of fish, wildlife, plants, or African elephant ivory made by him and the subsequent disposition made by him with respect to such fish, wildlife, plants, or ivory;

(B) at all reasonable times upon notice by a duly authorized representative of the Secretary, afford such representative access to his place of business, an opportunity to examine his inventory of imported fish, wildlife, plants, or African elephant ivory and the records required to be kept under subparagraph (A) of this paragraph, and to copy such records; and

(C) file such reports as the Secretary may require.

(3) REGULATIONS.—The Secretary shall prescribe such regulations as are necessary and appropriate to carry out the purposes of this subsection.

(4) RESTRICTION ON CONSIDERATION OF VALUE OR AMOUNT OF AFRICAN ELEPHANT IVORY IMPORTED OR EXPORTED.—In granting permission under this subsection for importation or exportation of African elephant ivory, the Secretary shall not vary the requirements for obtaining such permission on the basis of the value or amount of ivory imported or exported under such permission.

(e) REPORTS.—It is unlawful for any person importing or exporting fish or wildlife (other than shellfish and fishery products which (1) are not listed pursuant to section 4 of this Act as endangered or threatened species, and (2) are imported for purposes of human or animal consumption or taken in waters under the jurisdiction of the United States or on the high seas for recreational purposes) or plants to fail to file any declaration or report as the Secretary deems necessary to facilitate enforcement of this Act or to meet the obligations of the Convention.

(f) DESIGNATION OF PORTS.—(1) It is unlawful for any person subject to the jurisdiction of the United States to import into or export from the United States any fish or wildlife (other than shellfish and fishery products which (A) are not listed pursuant to section 4 of this Act as endangered species or threatened species, and (B) are imported for purposes of human or animal consumption or taken in waters under the jurisdiction of the United States or on the high seas for recreational purposes) or plants, except at a port or ports designated by the Secretary of the Interior. For the purposes of facilitating enforcement of this Act and reducing the costs thereof, the Secretary of the Interior, with approval of the Secretary of the Treasury and after notice and opportunity for public hearing, may, by regulation, designate ports and change such designations. The Secretary of the Interior, under such terms and conditions as he may prescribe, may permit the importation or exportation at nondesignated ports in the interest of the health or safety of the fish or wildlife or plants, or for other reasons if, in his discretion, he deems it appropriate and consistent with the purpose of this subsection.

(2) Any port designated by the Secretary of the Interior under the authority of section 4(d) of the Act of December 5, 1969 (16 U.S.C. 666cc–4(d), shall, if such designation is in effect on the day before the date of the enactment of this Act, be deemed to be a port designated by the Secretary under paragraph (1) of this subsection until such time as the Secretary otherwise provides.

(g) VIOLATIONS.—It is unlawful for any person subject to the jurisdiction of the United States to attempt to commit, solicit another to commit, or cause to be committed, any offense defined in this section.

EXCEPTIONS

SEC. 10. (a) PERMITS.—(1) The Secretary may permit, under such terms and conditions as he shall prescribe—

(A) any act otherwise prohibited by section 9 for scientific purposes or to enhance the propagation or survival of the affected species, including, but not limited to, acts necessary for the establishment and maintenance of experimental populations pursuant subsection (j); or

(B) any taking otherwise prohibited by section 9(a)(1)(B) if such taking is incidental to, and not the purpose of, the carrying out of an otherwise lawful activity.

(2)(A) No permit may be issued by the Secretary authorizing any taking referred to in paragraph (1)(B) unless the applicant therefor submits to the Secretary a conservation plan that specifies—

(i) the impact which will likely result from such taking;

(ii) what steps the applicant will take to minimize and mitigate such impacts, and the funding that will be available to implement such steps;

(iii) what alternative actions to such taking the applicant considered and the reasons why such alternatives are not being utilized; and

(iv) such other measures that the Secretary may require as being necessary or appropriate for purposes of the plan.

(B) If the Secretary finds, after opportunity for public comment, with respect to a permit application and the related conservation plan that—

(i) the taking will be incidental;

(ii) the applicant will, to the maximum extent practicable, minimize and mitigate the impacts of such taking;

(iii) the applicant will ensure that adequate funding for the plan will be provided;

(iv) the taking will not appreciably reduce the likelihood of the survival and recovery of the species in the wild; and

(v) the measures, if any, required under subparagraph (A)(iv) will be met;

and he has received such other assurances as he may require that the plan will be implemented, the Secretary shall issue the permit. The permit shall contain such terms and conditions as the Secretary deems necessary or appropriate to carry out the purposes of this paragraph, including, but not limited to, such reporting requirements as the Secretary deems necessary for determining whether such terms and conditions are being complied with.

(C) The Secretary shall revoke a permit issued under this paragraph if he finds that the permittee is not complying with the terms and conditions of the permit.

(b) HARDSHIP EXEMPTIONS.—(1) If any person enters into a contract with respect to a species of fish or wildlife or plant before the date of the publication in the Federal Register of notice of consideration of that species as an endangered species and the subsequent listing of that species as an endangered species pursuant to section 4 of this Act will cause undue hardship to such person under the contract, the Secretary, in order to minimize such hardship, may exempt such person from the application of section 9(a) of this Act to the extent the Secretary deems appropriate if such person applies to him for such exemption and includes with such application such information as the Secretary may require to prove such hardship; except that (A) no such exemption shall be for a duration of more than one year from the date of publication in the Federal Register of notice of consideration of the species concerned, or shall apply to a quantity of fish or wildlife or plants in excess of that specified by the Secretary; (B) the one-year period for those species of fish or wildlife listed by the Secretary as endangered prior to the effective date of this Act shall expire in accordance with the terms of section 3 of the Act of December 5, 1969 (83 Stat. 275); and (C) no such exemption may be granted for the importation or exportation of a specimen listed in Appendix I of the Convention which is to be used in a commercial activity.

(2) As used in this subsection, the term "undue economic hardship" shall include, but not be limited to:

(A) substantial economic loss resulting from inability caused by this Act to perform contracts with respect to species of fish and wildlife entered into prior to the date of publication in the Federal Register of a notice of consideration of such species as an endangered species;

(B) substantial economic loss to persons who, for the year prior to the notice of consideration of such species as an endangered species, derived a substantial portion of their income from the lawful taking of any listed species, which taking would be made unlawful under this Act; or

(C) curtailment of subsistence taking made unlawful under this Act by persons (i) not reasonably able to secure other sources of subsistence; and (ii) dependent to a substantial extent upon hunting and fishing for subsistence; and (iii) who must engage in such curtailed taking for subsistence purposes.

(3) The Secretary may make further requirements for a showing of undue economic hardship as he deems fit. Exceptions granted under this section may be limited by the Secretary in his discretion as to time, area, or other factor of applicability.

(c) NOTICE AND REVIEW.—The Secretary shall publish notice in the Federal Register of each application for an exemption or permit which is made under this section. Each notice shall invite the submission from interested parties, within thirty days after the date of the notice, of written data, views, or arguments with respect to the application; except that such thirty-day period may be waived by the Secretary in an emergency situation where the health or life of an endangered animal is threatened and no reasonable alternative is available to the applicant, but notice of any such waiver shall be published by the Secretary in the Federal Register within ten days following the issuance of the exemption or permit. Information received by the Secretary as part of any application shall be available to the public as a matter of public record at every stage of the proceeding.

(d) PERMIT AND EXEMPTION POLICY.—The Secretary may grant exceptions under subsections (a)(1)(A) and (b) of this section only if he finds and publishes his finding in the Federal Register that (1) such exceptions were applied for in good faith, (2) if granted and exercised will not operate to the disadvantage of such endangered species, and (3) will be consistent with the purposes and policy set forth in section 2 of this Act.

(e) ALASKA NATIVES.—(1) Except as provided in paragraph (4) of this subsection the provisions of this Act shall not apply with respect to the taking of any endangered species or threatened species, or the importation of any such species taken pursuant to this section, by—

(A) any Indian, Aleut, or Eskimo who is an Alaskan Native who resides in Alaska; or

(B) any non-native permanent resident of an Alaskan native village;

if such taking is primarily for subsistence purposes. Non-edible by-products of species taken pursuant to this section may be sold in interstate commerce when made into authentic native articles of handicrafts and clothing; except that the provisions of this subsection shall not apply to any non-native resident of an Alaskan native village found by the Secretary to be not primarily dependent upon the taking of fish and wildlife for consumption or for the creation and sale of authentic native articles of handicrafts and clothing.

(2) Any taking under this subsection may not be accomplished in a wasteful manner.

(3) As used in this subsection—

(i) The term "subsistence" includes selling any edible portion of fish or wildlife in native villages and towns in Alaska for native consumption within native villages or towns; and

(ii) The term "authentic native articles of handicrafts and clothing" means items composed wholly or in some significant respect to natural materials, and which are produced, decorated or fashioned in the exercise of traditional native handicrafts without the use of pantographs, multiple carvers, or other mass copying devices. Traditional native handicrafts include, but are not limited to, weaving, carving, stitching, sewing, lacing, beading, drawing, and painting.

(4) Notwithstanding the provisions of paragraph (1) of this subsection, whenever the Secretary determines that any species of fish or wildlife which is subject to taking under the provisions of this subsection is an endangered species or threatened species, and that such taking materially and negatively affects the threatened or endangered species, he may prescribe regulations upon the taking of such species by any such Indian, Aleut, Eskimo, or non-native Alaskan resident of an Alaskan native village. Such regulations may be established with reference to species, geographical description of the area included, the season for taking, or any other factors related to the reason for establishing such regulations and consistent with the policy of this Act. Such regulations shall be prescribed after a notice and hearings in the affected judicial districts of Alaska and as otherwise required by section 103 of the Marine Mammal Protection Act of 1972, and shall be removed as soon as

the Secretary determines that the need for their impositions has disappeared.

(f)(1) As used in this subsection—

(A) The term "pre-Act endangered species part" means—

(i) any sperm whale oil, including derivatives thereof, which was lawfully held within the United States on December 28, 1973, in the course of a commercial activity; or

(ii) any finished scrimshaw product, if such product or the raw material for such product was lawfully held within the United States on December 28, 1973, in the course of a commercial activity.

(B) The term "scrimshaw product" means any art form which involves the substantial etching or engraving of designs upon, or the substantial carving of figures, patterns, or designs from, any bone or tooth of any marine mammal of the order Cetacea. For purposes of this subsection, polishing or the adding of minor superficial markings does not constitute substantial etching, engraving, or carving.

(2) The Secretary, pursuant to the provisions of this subsection, may exempt, if such exemption is not in violation of the Convention, any pre-Act endangered species part from one or more of the following prohibitions:

(A) The prohibition on exportation from the United States set forth in section 9(a)(1)(A) of this Act.

(B) Any prohibition set forth in section 9(a)(1) (E) or (F) of this Act.

(3) Any person seeking an exemption described in paragraph (2) of this subsection shall make application therefor to the Secretary in such form and manner as he shall prescribe, but no such application may be considered by the Secretary unless the application—

(A) is received by the Secretary before the close of the one-year period beginning on the date on which regulations promulgated by the Secretary to carry out this subsection first take effect;

(B) contains a complete and detailed inventory of all pre-Act endangered species parts for which the applicant seeks exemption;

(C) is accompanied by such documentation as the Secretary may require to prove that any endangered species part or product claimed by the applicant to be a pre-Act endangered species part is in fact such a part; and

(D) contains such other information as the Secretary deems necessary and appropriate to carry out the purposes of this subsection.

(4) If the Secretary approves any application for exemption made under this subsection, he shall issue to the applicant a certificate of exemption which shall specify—

(A) any prohibition in section 9(a) of this Act which is exempted;

(B) the pre-Act endangered species parts to which the exemption applies;

(C) the period of time during which the exemption is in effect, but no exemption made under this subsection shall have force and effect after the close of the three-year period beginning on the date of issuance of the certificate unless such exemption is renewed under paragraph (8); and

(D) any term or condition prescribed pursuant to paragraph (5) (A) or (B), or both, which the Secretary deems necessary or appropriate.

(5) The Secretary shall prescribe such regulations as he deems necessary and appropriate to carry out the purposes of this subsection. Such regulations may set forth—

(A) terms and conditions which may be imposed on applicants for exemptions under this subsection (including, but not limited to, requirements that applicants register inventories, keep complete sales records, permit duly authorized agents of the Secretary to inspect such inventories and records, and periodically file appropriate reports with the Secretary); and

(B) terms and conditions which may be imposed on any subsequent purchaser of any pre-Act endangered species part covered by an exemption granted under this subsection;

to insure that any such part so exempted is adequately accounted for and not disposed of contrary to the provisions of this Act. No regulation prescribed by the Secretary to carry out the purposes of this subsection shall be subject to section 4(f)(2)(A)(i) of this Act.

(6)(A) Any contract for the sale of pre-Act endangered species parts which is entered into by the Administrator of General Services prior to the effective date of this subsection and pursuant to the notice published in the Federal Register on January 9, 1973, shall not be rendered invalid by virtue of the fact that fulfillment of such contract may be prohibited under section 9(a)(1)(F).

(B) In the event that this paragraph is held invalid, the validity of the remainder of the Act, including the remainder of this subsection, shall not be affected.

(7) Nothing in this subsection shall be construed to—

(A) exonerate any person from any act committed in violation of paragraphs (1)(A), (1)(E), or (1)(F) of section 9(a) prior to the date of enactment of this subsection; or

(B) immunize any person from prosecution for any such act.

(8)(A)(i) Any valid certificate of exemption which was renewed after October 13, 1982, and was in effect on March 31, 1988, shall be deemed to be renewed for a 6-month period beginning on the date of enactment of the Endangered Species Act Amendments of 1988. Any person holding such a certificate may apply to the Secretary for one additional renewal of such certificate for a period not to exceed 5 years beginning on the date of such enactment.

(B) If the Secretary approves any application for renewal of an exemption under this paragraph, he shall issue to the applicant a certificate of renewal of such exemption which shall provide that all terms, conditions, prohibitions, and other regulations made applicable by the previous certificate shall remain in effect during the period of the renewal.

(C) No exemption or renewal of such exemption made under this subsection shall have force and effect after the expiration date of the certificate of renewal of such exemption issued under this paragraph.

(D) No person may, after January 31, 1984, sell or offer for sale in interstate or foreign commerce, any pre-Act finished scrimshaw product unless such person holds a valid certificate of exemption issued by the Secretary under this subsection, and unless such product or the raw material for such product was held by such person on October 13, 1982.

(g) In connection with any action alleging a violation of section 9, any person claiming the benefit of any exemption or permit under this Act shall have the burden of proving that the exemption or permit is applicable, has been granted, and was valid and in force at the time of the alleged violation.

(h) CERTAIN ANTIQUE ARTICLES.—(1) Sections 4(d), 9(a), and 9(c) do not apply to any article which—

(A) is not less than 100 years of age;

(B) is composed in whole or in part of any endangered species or threatened species listed under section 4;

(C) has not been repaired or modified with any part of any such species on or after the date of the enactment of this Act; and

(D) is entered at a port designated under paragraph (3).

(2) Any person who wishes to import an article under the exception provided by this subsection shall submit to the customs officer concerned at the time of entry of the article such documentation as the Secretary of the Treasury, after consultation with the Secretary of the Interior, shall by regulation require as being necessary to establish that the article meets the requirements set forth in paragraph (1) (A), (B), and (C).

(3) The Secretary of the Treasury, after consultation with the Secretary of the Interior, shall designate one port within each customs region at which articles described in paragraph (1) (A), (B), and (C) must be entered into the customs territory of the United States.

(4) Any person who imported, after December 27, 1973, and on or before the date of the enactment of the Endangered Species Act Amendments of 1978, any article described in paragraph (1) which—

(A) was not repaired or modified after the date of importation with any part of any endangered species or threatened species listed under section 4;

(B) was forfeited to the United States before such date of the enactment, or is subject to forfeiture to the United States on such date of enactment, pursuant to the assessment of a civil penalty under section 11; and

(C) is in the custody of the United States on such date of enactment;

may, before the close of the one-year period beginning on such date of enactment make application to the Secretary for return of the article. Application shall be made in such form and manner, and contain such documentation, as the Secretary prescribes. If on the basis of any such application which is timely filed, the Secretary is satisfied that the requirements of this paragraph are met with respect to the article concerned, the Secretary shall return the article to the applicant and the importation of such article shall, on and after the date of return, be deemed to be a lawful importation under this Act.

(i) NONCOMMERCIAL TRANSSHIPMENTS.—Any importation into the United States of fish or wildlife shall, if—

(1) such fish or wildlife was lawfully taken and exported from the country of origin and country of reexport, if any;

(2) such fish or wildlife is in transit or transshipment through any place subject to the jurisdiction of the United States en route to a country where such fish or wildlife may be lawfully imported and received;

(3) the exporter or owner of such fish or wildlife gave explicit instructions not to ship such fish or wildlife through any place subject to the jurisdiction of the United States, or did all that could have reasonably been done to prevent transshipment, and the circumstances leading to the transshipment were beyond the exporter's or owner's control;

(4) the applicable requirements of the Convention have been satisfied; and

(5) such importation is not made in the course of a commercial activity,

be an importation not in violation of any provision of this Act or any regulation issued pursuant to this Act while such fish or wildlife remains in the control of the United States Customs Service.

(j) EXPERIMENTAL POPULATIONS.—(1) For purposes of this subsection, the term "experimental population" means any population (including any offspring arising solely therefrom) authorized by the Secretary for release under paragraph (2), but only when, and at such times as, the population is wholly separate geographically from nonexperimental populations of the same species.

(2)(A) The Secretary may authorize the release (and the related transportation) of any population (including eggs, propagules, or individuals) of an endangered species or a threatened species outside the current range of such species if the Secretary determines that such release will further the conservation of such species.

(B) Before authorizing the release of any population under subparagraph (A), the Secretary shall by regulation identify the population and determine, on the basis of the best available information, whether or not such population is essential to the continued existence of an endangered species or a threatened species.

(C) For the purposes of this Act, each member of an experimental population shall be treated as a threatened species; except that—

(i) solely for purposes of section 7 (other than subsection (a)(1) thereof), an experimental population determined under subparagraph (B) to be not essential to the continued existence of a species shall be treated, except when it occurs in an area within the National Wildlife Refuge System or the National Park System, as a species proposed to be listed under section 4; and

(ii) critical habitat shall not be designated under this Act for any experimental population determined under subparagraph (B) to be not essential to the continued existence of a species.

(3) The Secretary, with respect to populations of endangered species or threatened species that the Secretary authorized, before the date of the enactment of this subsection, for release in geographical areas separate from the other populations of such species, shall determine by regulation which of such populations are an experimen-

tal population for the purposes of this subsection and whether or not each is essential to the continued existence of an endangered species or a threatened species.

PENALTIES AND ENFORCEMENT

SEC. 11. (a) CIVIL PENALTIES.—(1) Any person who knowingly violates, and any person engaged in business as an importer or exporter of fish, wildlife, or plants who violates, any provision of this Act, or any provision of any permit or certificate issued hereunder, or of any regulation issued in order to implement subsection (a)(1)(A), (B), (C), (D), (E), or (F), (a)(2)(A), (B), (C), or (D), (c), (d) (other than regulation relating to recordkeeping or filing of reports), (f), or (g) of section 9 of this Act, may be assessed a civil penalty by the Secretary of not more than $25,000 for each violation. Any person who knowingly violates, and any person engaged in business as an importer or exporter of fish, wildlife, or plants who violates, any provision of any other regulation issued under this Act may be assessed a civil penalty by the Secretary of not more than $12,000 for each such violation. Any person who otherwise violates any provision of this Act, or any regulation, permit, or certificate issued hereunder, may be assessed a civil penalty by the Secretary of not more than $500 for each such violation. No penalty may be assessed under this subsection unless such person is given notice and opportunity for a hearing with respect to such violation. Each violation shall be a separate offense. Any such civil penalty may be remitted or mitigated by the Secretary. Upon any failure to pay a penalty assessed under this subsection, the Secretary may request the Attorney General to institute a civil action in a district court of the United States for any district in which such person is found, resides, or transacts business to collect the penalty and such court shall have jurisdiction to hear and decide any such action. The court shall hear such action on the record made before the Secretary and shall sustain his action if it is supported by substantial evidence on the record considered as a whole.

(2) Hearings held during proceedings for the assessment of civil penalties by paragraph (1) of this subsection shall be conducted in accordance with section 554 of title 5, United States Code. The Secretary may issue subpoenas for the attendance and testimony of witnesses and the production of relevant papers, books, and documents, and administer oaths. Witnesses summoned shall be paid the same fees and mileage that are paid to witnesses in the courts of the United States. In case of contumacy or refusal to obey a subpoena served upon any person pursuant to this paragraph, the district court of the United States for any district in which such person is found or resides or transacts business, upon application by the United States and after notice to such person, shall have jurisdiction to issue an order requiring such person to appear and give testimony before the Secretary or to appear and produce documents before the Secretary, or both, and any failure to obey such order of the court may be punished by such court as a contempt thereof.

(3) Notwithstanding any other provision of this Act, no civil penalty shall be imposed if it can be shown by a preponderance of the

evidence that the defendant committed an act based on a good faith belief that he was acting to protect himself or herself, a member of his or her family, or any other individual from bodily harm, from any endangered or threatened species.

(b) CRIMINAL VIOLATIONS.—(1) Any person who knowingly violates any provision of this Act, of any permit or certificate issued hereunder, or of any regulation issued in order to implement subsection (a)(1)(A), (B), (C), (D), (E), or (F); (a)(2)(A), (B), (C), or (D), (c), (d) (other than a regulation relating to recordkeeping, or filing of reports), (f), or (g) of section 9 of this Act shall, upon conviction, be fined not more than $50,000 or imprisoned for not more than one year, or both. Any person who knowingly violates any provision of any other regulation issued under this Act shall, upon conviction, be fined not more than $25,000 or imprisoned for not more than six months, or both.

(2) The head of any Federal agency which has issued a lease, license, permit, or other agreement authorizing a person to import or export fish, wildlife, or plants, or to operate a quarantine station for imported wildlife, or authorizing the use of Federal lands, including grazing of domestic livestock, to any person who is convicted of a criminal violation of this Act or any regulation, permit, or certificate issued hereunder may immediately modify, suspend, or revoke each lease, license, permit, or other agreement. The Secretary shall also suspend for a period of up to one year, or cancel, any Federal hunting or fishing permits or stamps issued to any person who is convicted of a criminal violation of any provision of this Act or any regulation, permit, or certificate issued hereunder. The United States shall not be liable for the payments of any compensation, reimbursement, or damages in connection with the modification, suspension, or revocation of any leases, licenses permits stamps, or other agreements pursuant to this section.

(3) Notwithstanding any other provision of this Act, it shall be a defense to prosecution under this subsection if the defendant committed the offense based on a good faith belief that he was acting to protect himself or herself, a member of his or her family, or any other individual, from bodily harm from any endangered or threatened species.

(c) DISTRICT COURT JURISDICTION.—The several district courts of the United States; including the courts enumerated in section 460 of title 28, United States Code, shall have jurisdiction over any actions arising under this Act. For the purpose of this Act, American Samoa shall be included within the judicial district of the District Court of the United States for the District of Hawaii.

(d) REWARDS AND CERTAIN INCIDENTAL EXPENSES.—The Secretary or the Secretary of the Treasury shall pay, from sums received as penalties, fines, or forfeitures of property for any violation of this chapter or any regulation issued hereunder (1) a reward to any person who furnishes information which leads to an arrest, a criminal conviction, civil penalty assessment, or forfeiture of property for any violation of this chapter or any regulation issued hereunder, and (2) the reasonable and necessary costs incurred by any person in providing temporary care for any fish, wildlife, or plant pending the disposition of any civil or criminal proceeding alleging a violation of this chapter with respect to that fish, wildlife, or plant. The amount of the reward, if any, is to be designated by the Secretary or the Secretary of the Treasury, as appropriate. Any officer or employee of the United States or any State or local government who furnishes information or renders service in the performance of his official duties is ineligible for payment under this subsection. Whenever the balance of sums received under this section and section 6(d) of the Act of November 16, 1981 (16 U.S.C. 3375(d)) as penalties or fines, or from forfeitures of property, exceed $500,000, the Secretary of the Treasury shall deposit an amount equal to such excess balance in the cooperative endangered species conservation fund established under section 6(i) of this Act.

(e) ENFORCEMENT.—(1) The provisions of this Act and any regulations or permits issued pursuant thereto shall be enforced by the Secretary, the Secretary of the Treasury, or the Secretary of the Department in which the Coast Guard is operating, or all such Secretaries. Each such Secretary may utilize by agreement, with or without reimbursement, the personnel, services, and facilities of any other Federal agency or any State agency for purposes of enforcing this Act.

(2) The judges of the district courts of the United States and the United States magistrates may within their respective jurisdictions, upon proper oath or affirmation showing probable cause, issue such warrants or other process as may be required for enforcement of this Act and any regulation issued thereunder.

(3) Any person authorized by the Secretary, the Secretary of the Treasury, or the Secretary of the Department in which the Coast Guard is operating, to enforce this Act may detain for inspection and inspect any package, crate, or other container, including its contents, and all accompanying documents, upon importation or exportation. Such persons may make arrests without a warrant for any violation of this Act if he has reasonable grounds to believe that the person to be arrested is committing the violation in his presence or view and may execute and serve any arrest warrant, search warrant, or other warrant or civil or criminal process issued by any officer or court of competent jurisdiction for enforcement of this Act. Such person so authorized may search and seize, with or without a warrant, as authorized by law. Any fish, wildlife, property, or item so seized shall be held by any person authorized by the Secretary, the Secretary of the Treasury, or the Secretary of the Department in which the Coast Guard is operating pending disposition of civil or criminal proceedings, or the institution of an action in rem for forfeiture of such fish, wildlife, property, or item pursuant to paragraph (4) of the subsection; except that the Secretary may, in lieu of holding such fish, wildlife, property, or item, permit the owner or consignee to post a bond or other surety satisfactory to the Secretary, but upon forfeiture of any such property to the United States, or the abandonment or waiver of any claim to any such property, it shall be disposed of (other than by sale to the general public) by the Secretary in such a manner, consistent with the purposes of this Act, as the Secretary shall by regulation prescribe.

(4)(A) All fish or wildlife or plants taken, possessed, sold, purchased, offered for sale or purchase, transported, delivered, received, carried, shipped, exported, or imported contrary to the provisions of this Act, any regulation made pursuant thereto, or any permit or certificate issued hereunder shall be subject to forfeiture to the United States.

(B) All guns, traps, nets, and other equipment, vessels, vehicles, aircraft, and other means of transportation used to aid the taking, possessing, selling, purchasing, offering for sale or purchase, transporting, delivering, receiving, carrying, shipping, exporting, or importing of any fish or wildlife or plants in violation of this Act, any regulation made pursuant thereto, or any permit or certificate issued thereunder shall be subject to forfeiture to the United States upon conviction of a criminal violation pursuant to section 11(b)(1) of this Act.

(5) All provisions of law relating to the seizure, forfeiture, and condemnation of a vessel for violation of the customs laws, the disposition of such vessel or the proceeds from the sale thereof, and the remission or mitigation of such forfeiture, shall apply to the seizures and forfeitures incurred, or alleged to have been incurred, under the provisions of this Act, insofar as such provisions of law are applicable and not inconsistent with the provisions of this Act; except that all powers, rights, and duties conferred or imposed by the customs laws upon any officer or employee of the Treasury Department shall, for the purposes of this Act, be exercised or performed by the Secretary or by such persons as he may designate.

(6) The Attorney General of the United States may seek to enjoin any person who is alleged to be in violation of any provision of this Act or regulation issued under authority thereof.

(f) REGULATIONS.—The Secretary, the Secretary of the Treasury, and the Secretary of the Department in which the Coast Guard is operating, are authorized to promulgate such regulations as may be appropriate to enforce this Act, and charge reasonable fees for expenses to the Government connected with permits or certificates authorized by this Act including processing applications and reasonable inspections, and with the transfer, board, handling, or storage of fish or wildlife or plants and evidentiary items seized and forfeited under this Act. All such fees collected pursuant to this subsection shall be deposited in the Treasury to the credit of the appropriation which is current and chargeable for the cost of furnishing the services. Appropriated funds may be expended pending reimbursement from parties in interest.

(g) CITIZEN SUITS.—(1) Except as provided in paragraph (2) of this subsection any person may commence a civil suit on his own behalf—

(A) to enjoin any person, including the United States and any other governmental instrumentality or agency (to the extent permitted by the eleventh amendment to the Constitution), who is alleged to be in violation of any provision of this Act or regulation issued under the authority thereof; or

(B) to compel the Secretary to apply, pursuant to section 6(g)(2)(B)(ii) of this Act, the prohibitions set forth in or authorized pursuant to section 4(d) or section 9(a)(1)(B) of this Act with respect to the taking of any resident endangered species or threatened species within any State; or

(C) against the Secretary where there is alleged a failure of the Secretary to perform any act or duty under section 4 which is not discretionary with the Secretary.

The district courts shall have jurisdiction, without regard to the amount in controversy or the citizenship of the parties, to enforce any such provision or regulation or to order the Secretary to perform such act or duty, as the case may be. In any civil suit commenced under subparagraph (B) the district court shall compel the Secretary to apply the prohibition sought if the court finds that the allegation that an emergency exists is supported by substantial evidence.

(2)(A) No action may be commenced under subparagraph (1)(A) of this section—

(i) prior to sixty days after written notice of the violation has been given to the Secretary, and to any alleged violator of any such provision or regulation;

(ii) if the Secretary has commenced action to impose a penalty pursuant to subsection (a) of this section; or

(iii) if the United States has commenced and is diligently prosecuting a criminal action in a court of the United States or a State to redress a violation of any such provision or regulation.

(B) No action may be commenced under subparagraph (1)(B) of this section—

(i) prior to sixty days after written notice has been given to the Secretary setting forth the reasons why an emergency is thought to exist with respect to an endangered species or a threatened species in the State concerned; or

(ii) if the Secretary has commenced and is diligently prosecuting action under section 6(g)(2)(B)(ii) of this Act to determine whether any such emergency exists.

(C) No action may be commenced under subparagraph (1)(C) of this section prior to sixty days after written notice has been given to the Secretary; except that such action may be brought immediately after such notification in the case of an action under this section respecting an emergency posing a significant risk to the well-being of any species of fish or wildlife or plants.

(3)(A) Any suit under this subsection may be brought in the judicial district in which the violation occurs.

(B) In any such suit under this subsection in which the United States is not a party, the Attorney General, at the request of the Secretary, may intervene on behalf of the United States as a matter of right.

(4) The court, in issuing any final order in any suit brought pursuant to paragraph (1) of this subsection, may award costs of litigation (including reasonable attorney and expert witness fees) to any party, whenever the court determines such award is appropriate.

(5) The injunctive relief provided by this subsection shall not restrict any right which any person (or class of persons) may have under any statute or common law to seek enforcement of any standard or limitation or to seek any other relief (including relief against the Secretary or a State agency).

(h) COORDINATION WITH OTHER LAWS.—The Secretary of Agriculture and the Secretary shall provide for appropriate coordination of the administration of this Act with the administration of the animal quarantine laws (21 U.S.C. 101–105, 111–135b, and 612–614) and section 306 of the Tariff Act of 1930 (19 U.S.C. 1306). Nothing in this Act or any amendment made by this Act shall be construed as superseding or limiting in any manner the functions of the Secretary of Agriculture under any other law relating to prohibited or restricted importations or possession of animals and other articles and no proceeding or determination under this Act shall preclude any proceeding or be considered determinative of any issue of fact or law in any proceeding under any Act administered by the Secretary of Agriculture. Nothing in this Act shall be construed as superseding or limiting in any manner the functions and responsibilities of the Secretary of the Treasury under the Tariff Act of 1930, including, without limitation, section 527 of that Act (19 U.S.C. 1527), relating to the importation of wildlife taken, killed, possessed, or exported to the United States in violation of the laws or regulations of a foreign country.

ENDANGERED PLANTS

SEC. 12. The Secretary of the Smithsonian Institution, in conjunction with other affected agencies, is authorized and directed to review (1) species of plants which are now or may become endangered, or threatened and (2) methods of adequately conserving such species, and to report to Congress, within one year after the date of the enactment of this Act, the results of such review including recommendations for new legislation or the amendment of existing legislation.

CONFORMING AMENDMENTS

SEC. 13. (a) Subsection 4(c) of the Act of October 15, 1966 (80 Stat. 928, 16 U.S.C. 668dd(c)), is further amended by revising the second sentence thereof to read as follows: "With the exception of endangered species and threatened species listed by the Secretary pursuant to section 4 of the Endangered Species Act of 1973 in States wherein a cooperative agreement does not exist pursuant to section 6(c) of that Act, nothing in this Act shall be construed to authorize the Secretary to control or regulate hunting or fishing of resident fish and wildlife on lands not within the system."

(b) Subsection 10(a) of the Migratory Bird Conservation Act (45 Stat. 1224, 16 U.S.C. 715i(a)) and subsection 401(a) of the Act of June 15, 1935 (49 Stat. 383, 16 U.S.C. 715s(a)) are each amended by striking out "threatened with extinction," and inserting in lieu thereof the following: "listed pursuant to section 4 of the Endangered Species Act of 1973 as endangered species or threatened species."

(c) Section 7(a)(1) of the Land and Water Conservation Fund Act of 1965 (16 U.S.C. 4601–9(a)(1)) is amended by striking out:

"THREATENED SPECIES.—For any national area which may be authorized for the preservation of species of fish or wildlife that are threatened with extinction." and inserting in lieu thereof the following:

"ENDANGERED SPECIES AND THREATENED SPECIES.—For lands, waters, or interests therein, the acquisition of which is authorized under section 5(a) of the Endangered Species Act of 1973, needed for the purpose of conserving endangered or threatened species of fish or wildlife or plants."

(d) The first sentence of section 2 of the Act of September 28, 1962, as amended (76 Stat. 653, 16 U.S.C. 460k–1), is amended to read as follows:

"The Secretary is authorized to acquire areas of land, or interests therein, which are suitable for—

"(1) incidental fish and wildlife-oriented recreational development;

"(2) the protection of natural resources;

"(3) the conservation of endangered species or threatened species listed by the Secretary pursuant to section 4 of the Endangered Species Act of 1973; or

"(4) carrying out two or more of the purposes set forth in paragraphs (1) through (3) of this section, and are adjacent to, or within, the said conservation areas, except that the acquisition of any land or interest therein pursuant to this section shall be accomplished only with such funds as may be appropriated therefor by the Congress or donated for such purposes, but such property shall not be acquired with funds obtained from the sale of Federal migratory bird hunting stamps."

(e) The Marine Mammal Protection Act of 1972 (16 U.S.C. 1361–1407) is amended—

(1) by striking out "Endangered Species Conservation Act of 1969" in section 3(1)(B) thereof and inserting in lieu thereof the following: "Endangered Species Act of 1973";

(2) by striking out "pursuant to the Endangered Species Conservation Act of 1969" in section 101(a)(3)(B) thereof and inserting in lieu thereof the following: "or threatened species pursuant to the Endangered Species Act of 1973";

(3) by striking out "endangered under the Endangered Species Conservation Act of 1969" in section 102(b)(3) thereof and inserting in lieu thereof the following: "an endangered species or threatened species pursuant to the Endangered Species Act of 1973"; and

(4) by striking out "of the Interior and revisions of the Endangered Species List, authorized by the Endangered Species Conservation Act of 1969," in section 202(a)(6) thereof and inserting in lieu thereof the following: "such revisions of the endangered species list and threatened species list published pursuant to section 4(c)(1) of the Endangered Species Act of 1973".

(f) Section 2(l) of the Federal Environmental Pesticide Control Act of 1972 (Public Law 92–516) is amended by striking out the words "by the Secretary of the Interior under Public Law 91–135" and inserting in lieu thereof the words "or threatened by the Secretary pursuant to the Endangered Species Act of 1973".

REPEALER

SEC. 14. The Endangered Species Conservation Act of 1969 (sections 1 through 3 of the Act of October 15, 1966, and sections 1 through 6 of the Act of December 5, 1969; 16 U.S.C. 668aa—668cc–6), is repealed.

AUTHORIZATION OF APPROPRIATIONS

SEC. 15. (a) IN GENERAL.—Except as provided in subsections (b), (c), and (d), there are authorized to be appropriated—

(1) not to exceed $35,000,000 for fiscal year 1988, $36,500,000 for fiscal year 1989, $38,000,000 for fiscal year 1990, $39,500,000 for fiscal year 1991, and $41,500,000 for fiscal year 1992 to enable the Department of the Interior to carry out such functions and responsibilities as it may have been given under this Act;

(2) not to exceed $5,750,000 for fiscal year 1988, $6,250,000 for each of fiscal years 1989 and 1990, and $6,750,000 for each of fiscal years 1991 and 1992 to enable the Department of Commerce to carry out such functions and responsibilities as it may have been given under this Act; and

(3) not to exceed $2,200,000 for fiscal year 1988, $2,400,000 for each of fiscal years 1989 and 1990, and $2,600,000 for each of fiscal years 1991 and 1992, to enable the Department of Agriculture to carry out its functions and responsibilities with respect to the enforcement of this Act and the Convention which pertain to the importation or exportation of plants.

(b) EXEMPTIONS FROM ACT.—There are authorized to be appropriated to the Secretary to assist him and the Endangered Species Committee in carrying out their functions under section 7 (e), (g) and (h) not to exceed $600,000 for each for fiscal years 1988, 1989, 1990, 1991, and 1992.

(c) CONVENTION IMPLEMENTATION.—There are authorized to be appropriated to the Department of the Interior for purposes of carrying out section 8A(e) not to exceed $400,000 for each of fiscal years 1988, 1989, and 1990, and $500,000 for each of fiscal years 1991 and 1992, and such sums shall remain available until expended.

EFFECTIVE DATE

SEC. 16. This Act shall take effect on the date of its enactment.

MARINE MAMMAL PROTECTION ACT OF 1972

SEC. 17. Except as otherwise provided in this Act, no provision of this Act shall take precedence over any more restrictive conflicting provision of the Marine Mammal Protection Act of 1972.

ANNUAL COST ANALYSIS BY THE FISH AND WILDLIFE SERVICE

SEC. 18. On or before January 15, 1990, and each January 15 thereafter, the Secretary of the Interior, acting through the Fish and Wildlife Service, shall submit to the Congress an annual report covering the preceding fiscal year which shall contain—

(1) an accounting on a species by species basis of all reasonably unidentifiable Federal expenditures made primarily for the conservation of endangered or threatened species pursuant to this Act; and

(2) an accounting on a species by species basis for all reasonably identifiable expenditures made primarily for the conservation of endangered or threatened species pursuant to this Act by states receiving grants under section 6.

Biographical Information on Committee Members

CHAIR

Michael T. Clegg is acting dean of the College of Natural and Agricultural Sciences and professor of genetics, University of California, Riverside. His research interests are the dynamics of multilocus genetic systems, genetic demography of plant populations, selection component analysis, biochemical separation in populations, and plant molecular evolution. Dr. Clegg has a BS (1969) and PhD (1972), genetics, University of California, Davis.

MEMBERS

Gardner M. Brown, Jr. is professor of economics and adjunct professor, Institute for Environmental Studies, University of Washington, Seattle. His areas of expertise include the economics of fisheries, ocean resources, wildlife and endangered species; natural resource damage assessments; and nonmarket valuation. He has an AB, Antioch College; MA and PhD, University of California, Berkeley.

William Y. Brown is principal, RCG/Hagler Bailly Inc., Arlington, Va. He is director of the Environmental Law Institute, Environmental and Energy Study Institute, Center for Marine Conservation, and U.S. Environmental Training Institute. His areas of expertise include the Endangered Species Act, population biology of sea birds, and environmental law. He has a BS,

University of Virginia; MAT, Johns Hopkins University; PhD, zoology, University of Hawaii; and JD, Harvard Law School.

William L. Fink is associate professor and associate curator of fishes, Museum of Zoology, University of Michigan, Ann Arbor. Formerly, he was assistant curator of fishes and assistant professor, Museum of Comparative Zoology, Harvard University. His areas of expertise include systematics of neotropical freshwater fishes, systematics of mesopelagic fishes, and systematic theory. Dr. Fink has a BS, University of Miami; MS, University of Southern Mississippi; and PhD, biology, George Washington University.

John Harte holds a joint professorship in Soil Science and Energy and Resources at the University of California, Berkeley. His areas of expertise include subalpine ecology, ecological impacts of climate change and acid precipitation, ecosystem modeling, and conservation biology. He has a BA, Harvard University; and PhD, physics, University of Wisconsin.

Oliver A. Houck is professor of law, Tulane University, New Orleans, La. His areas of expertise include environmental law, natural resources law, criminal law, toxics and hazardous wastes, and water resources and international enviromental law. He has written on the Endangered Species Act. He has a BA, Harvard University; and JD, Georgetown Law Center.

Michael Lynch is director, Ecology and Evolution Program, University of Oregon, Eugene. His areas of expertise include the integration of ecology, genetics, and evolution; the limits to phenotypic evolution; the neutral theory of evolution; conservation biology; evolution of life-history strategies; biology of parthenogenesis; the biology of aging; and quantitative genetics methodology. Dr. Lynch has a BS, St. Bonaventure University, New York; and PhD, University of Minnesota.

Lynn A. Maguire is an associate professor, Practice of Environmental Management, Duke University, Durham, N.C. Her research interests include the application of quantitative methods (including simulation modeling, statistics, and decision theory) to natural resource management; decision analysis and simulation modeling of endangered species management; conservation biology; forest fragmentation; and application of decision analysis for environmental dispute resolution in endangered species management and multiple-use land planning. She has an AB, Harvard University; MS, resource ecology, University of Michigan; and PhD, wildlife science, Utah State University.

Dennis D. Murphy is director, Center for Conservation Biology and senior research associate, Stanford University. His areas of expertise include habitat conservation planning, conservation biology, the Endangered Species Act, and the ecology of butterflies. Dr. Murphy has a BS, entomology, University of California, Berkeley; and PhD, biological sciences, Stanford University.

Patrick Y. O'Brien is team leader, Ecology, Chevron Research and Technology Company, Richmond, California. Dr. O'Brien has expertise in the following areas: environmental impact assessment, endangered species conservation planning, ecological risk assessment, wetlands delineation, natural resource damage assessment, habitat restoration, and the environmental elements of oil spill contingency planning and response. He has a BA, zoology, University of California, Berkeley; MS, water quality biology, University of California, Irvine; and PhD, ecology, University of California, Irvine.

Steward T. A. Pickett is member, Rutgers Graduate Ecology Faculty; adjunct professor, University of Connecticut; and scientist, Institute of Ecosystem Studies, Millbrook, New York. His areas of expertise include succession in plants, comparative ecology, and effects of disturbance on plant ecology. Dr. Pickett has a BS, University of Kentucky; and PhD, botany, University of Illinois, Urbana.

Katherine Ralls is a research zoologist, Smithsonian Institution, Washington, D.C. Her areas of expertise are the biology of mammals, mammalian social behavior, conservation biology, the genetic problems of small captive and wild populations, field studies of threatened and endangered species, and the development and testing of decision-making tools to improve management of threatened and endangered species. Dr. Ralls has a BA, Stanford University; MA, Radcliffe College; and PhD, biology, Harvard University.

Beryl B. Simpson is chair, Department of Botany and professor of botany, The University of Texas, Austin. Dr. Simpson's areas of expertise are plant ecology and evolution. Dr. Simpson has an AB, Radcliffe College; and MA and PhD, Harvard University.

Rollin D. Sparrowe is president, Wildlife Management Institute, Washington, D.C. His areas of expertise include research, legislation, and implementation of public programs that benefit wildlife; and development and implementation of federal policy on migratory birds, wetlands, waterfowl, migratory-bird hunting regulations, and endangered species. He has a BS, game management, Humboldt State University; MS, wildlife management, South Dakota State University; and PhD, wildlife ecology, Michigan State University.

David W. Steadman is senior scientist and curator of birds, New York State Museum and Biological Survey, and adjunct curator of fossil birds, Burke Memorial Museum, University of Washington. Dr. Steadman's areas of expertise include the systematics, biogeography, conservation, ecology, and paleoecology of vertebrates. He has a BS, biology, Edinboro State College; MS, zoology, University of Florida; and PhD, geosciences, University of Arizona.

James M. Sweeney is manager, Wildlife Issues, Champion International Corporation. His areas of expertise include habitat use and ecology of white-tailed deer and eastern wild turkey, elk, feral hogs, bobwhite quail, and spotted owls; stand dynamics of southern pine beetle infestations; oven bird and wood thrush, neotropical migratory birds, and other forestry-wildlife habitat interactions. Dr. Sweeney has a BS, forestry, MS, wildlife, University of Georgia; and PhD, wildlife, Colorado State University.

STAFF

David Policansky is associate director of the Board on Environmental Studies and Toxicology at the National Research Council, Washington, D.C. His interests include genetics, evolution, and ecology, particularly the effects of fishing on fish populations, ecological risk assessment, and natural resource management. He has a BA, biology, from Stanford University and MS and PhD, biology, from the University of Oregon.

Patricia Peacock, now director of partner programs at the Wildlife Management Institute, Washington, D.C., was staff officer, Board on Environmental Studies and Toxicology, National Research Council, Washington, D.C. (until February 1995). Her interests include the management of natural resources, especially forestry and fisheries. She has a BS, pharmacy, from the University of Montana, MS, fisheries, from the University of Alaska, and MPA from the Kennedy School of Government, Harvard University.

Adriénne Davis is senior program assistant in the Board on Environmental Studies and Toxicology, National Research Council, Washington, D.C. Her interests are information technology and management and education. She has a BS, business education, from the University of Maryland and MA, computers in education and training, from Trinity College.

Index

OTHER RECENT REPORTS OF
THE BOARD ON ENVIRONMENTAL STUDIES AND TOXICOLOGY

Wetlands: Characteristics and Boundaries (1995)

Biologic Markers in Urinary Toxicology (1995)

Review of EPA's Environmental Monitoring and Assessment Program (three reports, 1994-1995)

Science and Judgment in Risk Assessment (1994)

Ranking Hazardous Sites for Remedial Action (1994)

Pesticides in the Diets of Infants and Children (1993)

Issues in Risk Assessment (1993)

Setting Priorities for Land Conservation (1993)

Protecting Visibility in National Parks and Wilderness Areas (1993)

Biologic Markers in Immunotoxicology (1992)

Dolphins and the Tuna Industry (1992)

Environmental Neurotoxicology (1992)

Hazardous Materials on the Public Lands (1992)

Science and the National Parks (1992)

Animals as Sentinels of Environmental Health Hazards (1991)

Assessment of the U.S. Outer Continental Shelf Environmental Studies Program, Volumes I-IV (1991-1993)

Human Exposure Assessment for Airborne Pollutants (1991)

Monitoring Human Tissues for Toxic Substances (1991)

Rethinking the Ozone Problem in Urban and Regional Air Pollution (1991)

Decline of the Sea Turtles (1990)

Tracking Toxic Substances at Industrial Facilities (1990)

Biologic Markers in Pulmonary Toxicology (1989)

Biologic Markers in Reproductive Toxicology (1989)

Copies of these reports may be ordered from
the National Academy Press
(800) 624-6242
(202) 334-3313